RuMBLings:
A History and Highlights of The Rocky Mountain Biological Laboratory

By
Ellen F. Dobbin

RuMBLings: A History and Highlights of the
Rocky Mountain Biological Laboratory

Contact dobbinbooks@gmail.com

Contents

The Science

Acknowledgements

About the Artist

Works Cited

Bibliography

Foreward

Ellen Dobbin has given us a great gift with this engaging history and at the same time gotten to the root of why the Rocky Mountain Biological Laboratory has taken its place as a premiere biological field station and a world leader in biological research and education.

Of course, the foundation of this remarkable institution is the breathtaking landscape, first seen by RMBL founder Dr. John C. Johnson in 1919. He recognized immediately that it was a "paradise for the biologist." But the terrain alone, even such an exquisite setting, could not have accounted for RMBL's survival through its early years or its ability to thrive over the long term.

The true basis of RMBL's success has been the *exceptional people* who have graced its history and continue to be drawn to it.

Ellen grasped this fact from the beginning, telling RMBL's compelling story by introducing the reader to many of the unforgettable personalities who have shaped and guided RMBL and who have in turn become part of its very fabric.

RMBL has been a magnet for remarkable people from its earliest days, with each generation exerting its own powerful magnetic pull and developing deep personal friendships and fruitful professional affiliations. What are the common denominators, the qualities of so many diverse and distinctive personalities over so many years that collectively have helped RMBL take its place as a world-class institution? As one who has had the great fortune to know many of the people you will meet in this book, I can attest that many an entertaining evening could be spent ruminating on that question.

There are overarching qualities that describe most of the personalities who are, or have been, a part of what has grown into RMBL's global yet surprisingly close-knit community: an intense intellectual curiosity, an unabashed enthusiasm for the natural world, solid competence in a chosen field, an appreciation for colleagues and community, and a belief that what RMBL makes possible really *matters* in the world.

Here are the stories of many exceptional individuals who, like countless others, whether they have experienced Gothic for a day or a decade, have been captivated by RMBL and contributed to its inspiring success. A most enjoyable and satisfying read.

Carol M. Johnson
Gunnison, Colorado and Amherst, Massachusetts
May, 2015

Carol is the granddaughter of RMBL founder John C. Johnson and began coming to Gothic in the early 1950s with her parents, Chris and Dorothy Johnson, who served RMBL for over 50 years and gave the institution their all throughout their lives. Carol and her brothers Doug and Dan spent childhood summers growing up at Gothic. As young adults, the three worked on RMBL's kitchen, office and facilities crews before pursuing careers in law, science teaching and medicine, all the while maintaining close ties with the RMBL community. Carol met her biologist husband at RMBL and worked several summers as his field assistant. In recent years, Carol has also served on the RMBL Board of Trustees.

Preface

Weather is not an infrequent trigger for memories. The weather in the mountains of Colorado in August might be wet, or it might be dry – but there is always the possibility of rain. So it was the morning I first visited Gothic. The guest of some friends who have become deeply committed to the Rocky Mountain Biological Laboratory (RMBL), I tagged along as they and their grandchildren received a personal tour in August of 2010 by the Director of RMBL, Dr. Ian Billick. The ground was wet with morning dew and the sky threatened eventual showers, but we suited up with rain gear and sauntered along the trails to the parts of Gothic that might pique an outsider's interest. Here is the Barclay cabin, with its low ceilings, custom built by people barely five feet tall. What about that chicken coop, made a home for and by the Langenheims when there was insufficient money to build a proper cabin; it, unfortunately, no longer exists. Critters abound on the paths – a chipmunk. Yawn. No, a golden mantled ground squirrel! What is that? Keep to the paths, please – one might unknowingly crush a lovely wildflower or burying beetle, or step into an investigator's research area. Oh, and yes, the marmots! Baby marmots! On display for the benefit of the grandchildren – but aren't we all children when it comes to loving baby animals?

Throughout the walk, Dr. Billick kept up a continuing narrative about the background of the buildings, the people who built them, the town and the organization itself. It was fascinating to me, a relative newcomer to that part of the world and a "humanist", as opposed to "scientist". Quite frankly, I knew almost nothing about science and had no understanding of the work that had been going on in Gothic since 1928, and barely a sense that scientists are humanists, too. But the stories were good - too good to be lost. So… curious, I asked if anyone was writing them down. No, said the concerned Director, but "we are looking for an historian. Do you know anyone?" A long pause and a deep breath later, I raised my hand. And that was the beginning of a journey that has introduced me to some of the most fascinating individuals I have ever met and educated me (somewhat) in the diverse and significant work they do.

The tiny ghost town of Gothic, Colorado nestled in the shadow of massive Gothic Mountain, is just three miles up the road from the ski resort town of Mt. Crested Butte, which in turn is just three miles up the road from the old mining town of Crested Butte. All these places, as is common, are so much more than just ski town or old mining town or ghost town. The mines no longer operate. The ski industry has been joined by ever more popular summer activities. Gothic isn't a ghost town anymore – and hasn't been since 1928. True, some months of the year see empty condos, deserted cabins, shuttered laboratories. Restaurants and inns in the resort towns take a well-earned break at times when the flowers aren't blooming, the snow isn't falling. Winter caretakers (and billy barr) are solitary in their oversight in Gothic. Mother Nature closes in around Her own. But time cycles back to busy seasons for skiers, hikers, bikers, outdoor enthusiasts of all ilk – and, scientists return to Gothic, to pick up where they left off, or perhaps begin anew.

Did it ever rain that day in August 2010? I can't recall, but the smell of rain was in the air, the dampness lending strength to the *frisson* I felt. Ian and I agreed to talk at a later date, so I could start asking questions that had yet to occur to me. Admittedly, trepidation set in. History imposes great responsibility upon those who choose to record it. Facts are lost in the mists of time. Conclusions must be drawn from minimum input, to use a 21st century term. The founders had all passed on; the second generation was on its way to joining them. The "archives" barely existed. Where was the information to be found?

People, that's where. This is not an oral history, but it is a collaborative work. It has never ceased to amaze me how willing the busy individuals of Gothic were to sit with me in their cabins, or in a coffee shop, or over lunch, or – most frequently – on the benches of the town library. It rained often – that I DO recall – and the winds blew dust around, or the sun warmed us slightly as the days grew cool. Coming from a long day of field work, most of them looked at me first with confusion, probably wondering why I didn't just go to their websites. I did. But one cannot get a true sense of a person from a website, nor can one ask questions, or request that the work be explained in everyday language. Gradually, they seemed to understand. This person, totally ignorant of what we do, is hoping to make it interesting and understandable to the lay reader. Gothic may be growing, but it must not outgrow its past.

This history is one of people and the work they have done and continue to do at the Rocky Mountain Biological Laboratory. Four summers were spent interviewing individuals, reinforced by many hours spent learning about the work in summer workshops and by simple, traditional research. Ian Billick opened what existed of the RMBL archives at the time to my prying eyes. Johnson family documents were put at my disposal. Yes, there is much that is not in this book - records of meetings of the Boards of Trustees, financial accountings, technical details of the operation of the organization. What I have endeavored to reveal is the heart of the place – to create a story that is a history – to make it interesting to those who know nothing about RMBL, as well as to those who think they know everything. It is for those of us who think we don't know "science". It is a chance to learn.

Ellen Dobbin

"Gothic"

- Anything relating to the Goths, their civilization or their language; Teutonic; Germanic; uncouth; barbarous
- A style of architecture developed in 12th century France
- A style of literature that uses desolate settings and includes mysterious or macabre events
- A mountain in the Elk Mountain range of central Colorado
- A small, formerly deserted mining town located in the shadow of Gothic Mountain, now the home of the

Rocky Mountain Biological Laboratory

Gunnison and Gothic

In an era that now seems like "once upon a time", high in the West Elk Mountains of the Colorado Rockies, there sprang up a boom-hopeful mining town which would come to be known as "Gothic" in honor of the mountain that overshadowed it. The year was 1879 and the location was approximately forty miles from the town of Gunnison. Exquisitely beautiful in summer, Gothic was transformed into an inaccessible, snow-bound hideaway by winter. Bitter cold prevailed. Fall and spring were "mud season", made tolerable by the eruption of wildflowers in spring and the golden glow of aspens in fall. Unless, of course, all one was interested in was silver – the reason for Gothic's existence.

The Place

Gunnison and the country surrounding it were, in the middle of the nineteenth century, the epitome of the Wild West. The area was so violent that "fur trappers virtually ignored it; so rugged that railroads skirted parts of it and failed in others; so tough that the Ute Indians moved out and spent their winters elsewhere".[1] An era that would be marked by gold fever saw this part of western Colorado overlooked for years for lack of the precious ore. Named for John Gunnison, who had come to the area in 1853 with the Army Topographical Engineers to survey it for a possible future railroad, the Gunnison Town Company was organized in 1879, the same year as the town of Gothic and only a year before Crested Butte, both farther up valley in the mountains.[2]

Of interest and pertinence to our tale are reactions from travelers and settlers during the early, boom times, even if their sense of geography may not always match that of today. The physical beauty of the area is undisputed. The approach to Gunnison from the east was described as, "a most beautiful valley, overlooking a stream of clear, crystal, bubbling water... The grass was green and full of wildflowers – such beauty of nature can never be told".[3]

Decades later, Gunnison itself was described as looking like a military camp, filled with tents and rough looking men, dressed in buckskin, carrying guns and chewing tobacco. Lodging was nearly non-existent, and "proper" ladies were conspicuous by their absence. One female traveler was allotted a sofa to sleep on in the town's hotel for one dollar a night, as there were no rooms available. She was admonished by the landlady to not undress and to sleep on top of any valuables. She records that she was unable to sleep, due to the continual, loud noise. " I have never heard worse. Stamping up and down the stairs… music going on, on [*sic*] both sides of the house, and the gamblers yelling their games out, they being in tents, was as bad as being outdoors. I had heard of Bedlam, but this beat that place."[4]

SILVER!

1879 was the year everything changed for the isolated, mountainous region and it was Gothic, not Gunnison, which became the "biggest 'boom' town of the north country".[5] Two old prospectors searching for silver in the Colorado high country, John and David Jennings, were financially backed, or "grubstaked", by a wealthy, Chicago hotel owner. This financier, Obediah Sands, was to receive a half interest in anything the brothers discovered – and they discovered silver. In May of 1879, word got out that silver had been found near what would become Gothic, Colorado, in the West Elk mountains, high above Copper Creek. "A flood of miners" mobbed the area, set up tents and went about looking for their own stakes. The confluence of Copper Creek and the East River hummed with activity and excitement; thus was born the little town of Gothic.[6] And what a town it was to become!

It is estimated that between two thousand and twenty thousand people headed to the Gunnison country in 1879, when word of the silver strike got out.[7] Located in a beautiful but isolated area of the Colorado mountains, Gothic was incorporated as a town on July 17, 1879. A place that in May of that year had been a quiet retreat for numerous wildlife and possibly a few Ute Indians became, by the end of the year, a bustling, vibrant community. Nearly two hundred buildings were completed; somewhere between five hundred and a thousand people called the town home and two silver smelters were partially completed. The *Saguache Chronicle* on August 2, 1979 wrote that, "Gothic City and Crested Butte are already putting on metropolitan airs, and the whole country is alive with people who have every confidence in the richness of the mines."[8]

The weather, alas, was not kind to the newcomers. The winter of 1879/80 was recorded as one of the worst ever in the area. At the end of the winter, only twenty families remained. "The snow was over six feet on the level and the only contact the little band had with the outside world came via Louis Barthell, the mail carrier", who daily traveled eight miles along the East River on skis between Gothic and Crested Butte.[9] According to the Gunnison newspaper of the day, "...Barthell once carried a five gallon can of oil and a fifty pound sack of flour along with his regular mail bag from Crested Butte to Gothic".[10]

Yet, miners were a hearty stock and, come spring, they returned in droves. Getting there was no easy task. Today's dirt road through Gothic would have been considered a major highway by the standards of the time. To reach the 9,485 foot town in the shadow of 12,635 foot Gothic Mountain, one had to push through quagmires of mud and muck in the spring, dry dust and rocks in summers, and slick ice in the late fall. Winters were out of the question – heavy snows obstructed any sign of a "road" and skis were the only mode of transportation, assuming one knew where to go. For better or for worse, those characteristics are not much changed today.

Toll roads were built by the enterprising. A buck was a buck, any way you could get away with making one! And the true-believing miners, dreaming incessantly of success, were willing to pay, if only they could hit pay dirt. The direct road from Crested Butte was not the only way to reach the mountainous area. "Especially dangerous was the feared Schofield to Gothic Toll Road, which was used in the early 1880s. From the top of Schofield Pass at 10,722 feet, one descended to a precipitous road which hung suspended on the side of a cliff, past Emerald Lake... [It] was ill-kept and had no turnouts or passing lanes... very steep, [it was] only seven to eight feet wide, instead of the required ten feet."[11] One had always to be on the lookout for mudslides, flash floods, rockslides, avalanches, icy patches and snow. Trusty burros, the pack animals of choice, formed trains loaded with mining tools and provisions for the miners. It was slow going, indeed.

By the summer of 1880, the official census recorded a permanent population in Gothic of 950 persons, which did not include the several thousands of transient miners. Included in the statistics were five law firms, four grocery stores, three restaurants, two hotels and "a disproportionate number of saloons". The Olds House Hotel had been built in 1879 and the Gothic Hotel sprung up in 1880. The town boasted its own newspaper, *The Elk Mountain Bonanza*, the fourth paper to be published in Gunnison County. A couple years later, the *Silver Record* began publication and lasted for about five years.[12] In addition, newspaper ads of the time revealed the existence of seven assaying/surveying firms; five real estate agencies; three drug stores, three doctors, two general merchandise stores, two blacksmiths, two boot/shoe makers, a butcher, a bank, a barber shop, one or two livery stables, a stationery store, a jeweler/watchmaker and more.[13] Gothic had all the indications of being a thriving town.

Conflicting tales of life in Gothic survive. One proud, enthusiastic resident boasted in a letter to the *Rocky Mountain News* at the end 1879 that, "We have no gambling dens here and only one saloon, which facts certainly speak well for the morals of our town, and the society here can be compared favorably with that of much older towns in the eastern states".[14] A different picture is painted by W.W. McKee, who would build a life in Gunnison as a successful business man. He describes Gothic as offering many saloons, two dance halls, numerous gambling places, Sunday horse races and dance hall girls in shockingly short skirts! And, the "women, if women there are, are fierce and restless, like the men".[15]

Adding to the excitement, in July of 1880 Barlow and Sanderson extended their stagecoach lines through Gunnison and the surrounding country. By the beginning of 1881, their coaches were connecting and transiting every major mining camp in the region. Every day, two lines carrying 50-60 people each way sailed behind heaving horses into Gothic, Crested Butte and Ruby to the north.[16] And several years later, in 1886, sleigh travel was inaugurated between Crested Butte and Aspen, via Gothic and East Maroon Pass. The sleighs, according the *Gunnison Daily Review Press,* were "new, elegantly upholstered, four-seated double boot bobs of the finest make, liberally furnished with robes and raps [*sic*] to make the trip not only comfortable but pleasant".[17]

Meanwhile, only a few miles down valley from Gothic, the mining town of Crested Butte was struggling to grow roots. The miners congregated there were searching not just for silver, but for copper and coal and anything else they could pull from the mountains. Situated at 8,800 feet in a small vale, it was surrounded by peaks with names like Whetstone, Avery, Gothic, Bellview, Teocali, Treasury, and Mt. Emmons, more commonly known as Red Lady. Now here is a tale; although it already had a perfectly respectable name, why did Mt. Emmons become commonly known as "Red Lady"? There are multiple versions of the story, as so often happens when truth is hidden in oral renderings and the passage of time. Perhaps the most dearly recalled and frequently repeated is this: That the red lights of the houses of ill repute in Crested Butte used to shine on the mountain, giving it a red glow, ergo Red Lady. This, of course, presupposes there were a sufficient number of such establishments in the town of Crested Butte to cast enough light to impact the color a mountain.

Aside from the silver, perhaps the most singular event to occur in little Gothic, Colorado was a visit by former United States President Ulysses S. Grant in 1880. As the story goes, former President Grant and his family had concluded a trip around the world in December of 1879, after which he toured cities and towns throughout America. While traveling around the state of Colorado with the governor and a small party he expressed a desire to see "the wildest mining camp in Colorado".[18] Having been told that Gothic was just the place, he and his entourage descended upon the town for two nights. They were escorted into town by a brass band, met with "a blast from the town's supply of giant powder", and entertained with a gala reception and ball.[19] They stayed at either the Olds Hotel or the newer Gothic Hotel, depending on one's source, and journeyed into the mountains to visit several mines.[20] Quite an honor for a town just entering its second year of life.

As the saying goes, however, all good things must come to an end – and Gothic's ending was all too soon in coming.

Hopeful miners continued to journey to Gothic and the West Elk Mountains until midway through the 1890s, but boom times ended rapidly. The grade of silver being pulled from the mines was not the highest and required processing. The Avery Smelter in Gothic was not placed into operation until the end of July 1881 and before that time all ore had to be transported to Gunnison for smelting and processing, which was expensive and time consuming. According to the *Gothic Miner*, "… the one great requirement of the Gothic… area was a smelting technique which was much cheaper than the ones being used by smelters at that time".[21] Additionally, circumstances in the mining world of Colorado contributed to the decline and eventual demise that would affect Gothic. Mines were staked and sold but not developed; when money was needed by purchasers, there was nothing to pull out of the undeveloped mines. And, always a favorite pastime in America, litigation over ownership tied up much of the land and many of the mines in courts. Therefore, they lay fallow.[22]

Not all was lost, for a while. Even in 1886 there were seven mines continuing to operate in the Gothic area. The original big strike, the Silvanite Mine, was profitable. Miners continued to search the area and work what they found, but by 1886 there wasn't much left of the little-town-that-was. Many businesses were closed or relocated, some to Gunnison. A traveler in the area commented as early as the summer of 1883 that, "Two years ago Gothic was larger and more flourishing than Crested Butte. To-day [*sic*] Gothic is dead, and Crested Butte thrives and grows".[23]

The glory days may have passed, but dreams die hard and hope lives eternal. Or so it is said. Our female traveler in 1883 who took pride in reporting on the area tells of conversations she had with men she encountered wandering through the mountains. One told her that, "there's plenty of good properties here, but the people are too poor to work them… everybody thought that mountain [Gothic] was made of solid silver; and so 't is, pretty near, if there was only any getting at it." A little farther along, she met another Gothic man, "rosy, hearty, accoutered in fringed buckskin, with a canopy-brimmed yellow sombrero, and he galloped along as if he owned the earth and the air." Asking him his opinion of the area and its future, he retorted that it was "full of splendid properties" and would be a "first-rate camp to live in when they got things fixed up a little".[24]

A slightly less enthusiastic summary was offered by a woman in nearby Schofield. "Waiting for a boom, that's what's the matter with this town. I've got no patience with this boom business. It's the ruination of this country. It just spoils everything. There isn't a decent house in the town, and there won't never be." The landlord of the "board shanty" – or "inn" – in Schofield reported on the tenth of August that the camp had been pretty dull and they hadn't gotten their horses in yet. "You see, there was foot of snow lyin' in the street here the 22nd of June and that's put things back… but they're comin' now, fast." In only a few short weeks, the snowflakes would be flying again.[25]

So little Gothic, Colorado essentially ceased to be. It was nearly deserted, its buildings allowed to fall into disrepair and ruin. Mother Nature returned to do what she does best – reclaim that which is hers – except, of course, for those bits and pieces of her domain that man would not relinquish. Enter Garwood Hall Judd, the man who never left or, as he is better known, "The Man Who Stayed".

Garwood Hall Judd — "The Man Who Stayed" (1852-1930)

The word "history" contains within its core the word "story". History itself, though honorable and even venerable, has been not infrequently rendered dull, dry and downright boring by the insistence of the historian that one must relate the facts, ma'am, just the facts. Facts are the foundation of true rendition, but anyone with an iota of intelligence, lively curiosity and an appreciation for the humorous, the sordid, the lascivious, the spicy, the good, the bad, the beautiful, the ugly, knows that facts alone do not a tale make. And we all do love a good tale, especially a tall one. Perhaps just slightly embellished - just to make it memorable, legendary. We do love our heroes and heroines. We love our "stories".

The history, or story, of the American West is filled with many celebrated tales, making it the stuff of legends. The people who populated the times of the emerging west are some of the most famous and beloved in American folklore. Who does not recognize the names and know a bit about such individuals as Wild Bill Hickok, Jesse James, Doc Holliday? Or perhaps the towns of Tombstone, Santa Fe, Kansas City? And, of course, one cannot discount Sitting Bull, Crazy Horse, the Navajo, the Ute, the Cheyenne... Stories of lands lost and won, hunting grounds usurped for farming, Wild West shows to make everyone feel good about it all. White men, red men, horses, wagons, bandits, schoolmarms, scalpings, warfare - but all so terribly romantic!

Into our particular story enters the first of several Gothic, Colorado legends, a man by the name of Garwood Hall Judd. He was born in Norwalk, Ohio on March 14, 1852 to Erasmus Darwin Judd, a sixth generation American whose roots traced back to early 17th century Connecticut, and Phoebe Hall, also from Connecticut. He died on May 15, 1930 in Crested Butte, Colorado, the first of his family to be "buried" west of the Continental Divide.[1] According to local historians, Judd's wishes were respected after his death; his body was cremated and the ashes spread around the town he had made his home – Gothic.

There is not much recorded about Judd, but he did keep a personal journal for the year 1880, when he left Kansas and traveled to Colorado. This journal was carefully preserved and safely secured after Judd died in 1930 – and, even better, it somewhat mysteriously resurfaced in 2013.[2]

According to his own record, Judd arrived in Gothic on Wednesday, May 25, 1880, at 1:00 in the afternoon. He was 28 years old and seeking, as did many others, to make his fortune in silver. He was unusually well educated for his time and among his peers in Gothic, having attended Oberlin College in Oberlin, Ohio.[3] Or was it Knox College, in Galesburg, Illinois?[4] It depends on who is telling the tale.

Conjecture had it for many years that he was either not suited to mining, physically or temperamentally, or perhaps simply did not like it. This is borne out by entries in his journal during May 1880, after his arrival in Gunnison but before he went to Gothic. His spirits are frequently depressed and he writes often of sickness and cold and loneliness.

The diary begins in Irving, Kansas on January 1, 1880… Judd appears to be happy and speaks of going dancing with his "girl", Ada E. Clark, an elementary school teacher. He appears to be unemployed and generally not engaged in any money-making enterprise, but never talks of being short of funds. On January 5, his entry states, "My thoughts today are something like this – have been telling everybody I was going West in order to make some collections. Must go or they will not respect my word in future." These are not the words of an enthusiastic prospector or fortune seeker. We are never enlightened about exactly what those "collections" were. He does seem to be struggling with some personal demons, speaking of "contraband" and "dissipation" and "too many cigars" and at one point makes reference to "the State of Kansas vs. Garwood Hall Judd" on March 9, 1878. He was a bit of a rogue, recording for posterity his attempted romantic advances toward women other than the one he calls his "girl", and jotting down that he thought it was "good logic" to follow the adage, "When one has not the one whom one loves, one must love the one that one has". Ada Clark did not agree, telling him it was "all wrong" and he "mustn't do it".

On Monday, January 19th, 1880, at 4:48 pm, Judd departed Irving, Kansas for the town of Waterville, Kansas. This is the first of many stops which would bring him back to Irving, and then on to Colorado. "I know I have all my morals with me and I left no creditors behind. Here I go…" He was seen off by some friends, noting "I thank God I've made two good little friends in six years". Apparently Ada accompanied him for the first night of his journey as he mentions they "took in Kendall play", which he did not think was very good.[5] The next day, he bid Ada goodbye and took the 3:15 pm train for Clifton, also in Kansas. He was not happy about his decision, writing that he was "feeling very bad. If I had a home would go there. I guess just as well here as anywhere".

Judd traveled by train (and apparently on foot) through western Kansas, passing through and staying in various towns on his way to

Colorado, occasionally backtracking. Along the way, he suffered physical illnesses and bouts of serious homesickness. He was not alone. A variety of friends crop up repeatedly in his jottings, and several of them would accompany him all the way to Gunnison and beyond to Gothic. Some he refers to by full names, others by initials only. Some get along with everyone, others with only some and only sometimes. He records drinking bouts, fights, stabbings, arson, court trials, vagrants getting thrown off trains - and constantly looking for letters from Ada. On Thursday, February 12 he mentions the "first fun of the trip tonight. Comfort and I serenaded our new biscuit shooter", an old western term for a cook or waiter in a camp. Judging from the entries in his diary, Judd did like to sing.

He also liked a good, raucous joke played against a friend. In Gaylord, Kansas on February 23rd they bought a cow and a yoke of steers. His friend, Wallace Comfort, "is the boss, quite a pompous sort of fellow not used to cow driving. As I got him to ride (or rather try to) a cow across the Solomon @ Gaylord, Pratte Mills… cow bucked and ran over him." Judd was so tickled by this that he felt compelled to comment – "Godfrey damn! What fun today!" He not infrequently mentions buying cows, but never says what he does with them. Most likely, he has them driven (by Comfort) to stockyards to be sold. It is clear that entertaining themselves during the long days of traveling seems to have been important to the men on their way to try to make fortunes.

April 10th was the day Judd arrived in Colorado, in Pueblo to be precise. He negotiated with the landlady at the Victoria Hotel for himself and his four traveling companions for rooms at $1.35 a day or $7.00 a week. The weekly rate turned out to be unnecessary as two days later they continued on to Canon City, where they learned that there were nine feet of snow at Marshall Pass, six in Poncha. Undeterred, these errant prospectors booked passage for Gunnison, "a trip to be made in fifteen days". Their first night out, when they made camp, they discovered that in their rush to depart not all the provisions had made it onto the wagons. It was "found we had forgotten onions and other spices – all but the whiskey. Seems we had all remembered it."

The journey over the mountains and across the continental divide was a challenging one. Snow, mush for supper, deeper snow and more mush with "flippers" [possibly a kind of hotcake disliked by Judd] defined their days and meals. The hot springs near Poncha were welcome for their warmth. As they climbed higher, the thin air "affected us all more or less, throats sore. Had whiskey, canned beef and pie" for supper. Bread was a luxury, which they were able to obtain only in Poncha. But by the time they arrived in Saguache on April 21st, nine days into their trek, Judd was able to report "roast beef, potatoes, custard pie" for dinner. Food mattered to these hungry wayfarers.

Sunday, April 25th saw the motley crew pulling into the "new Jerusalem, Gunnison City. Stopped and looked all round for the town. Saw the usual fifty buildings, large and small. People don't stop for today – would but it was the middle of the week. Snowing hard. Will not pass judgment on town until I see more." They settled into camp three-quarters of a mile from town, by the Gunnison River. Baths were seen to, mail was collected, meals were prepared and consumed with gusto. Inquiries into the surrounding mining camps were made – five feet of snow was reported at Ruby Camp, where lots were selling sight unseen for between $40 and $200. Crested Butte was said to be not so expensive. But Crested Butte did not boast having any silver.

The "Silver King" mine and "Democrat Lode", which were goals of the group, were located on April 29th and May 1st. "Bouyant" feelings gave way to disappointment as the mines were found to be worthless, "even the iron all rotted out of her". The weather continued to be cold and snowy, icy winds blew incessantly, illness moved from tent to tent and the occasional warmer day was embraced and enjoyed. Interestingly, although Gunnison was known to be a "wild west town", Judd refrained from commenting on that aspect of it, with one exception. On May 11th he notes that the "Lake City Silverworld reports party of 25 miners corralled and fighting Utes on Indian Creek, 35 miles S West of here. Going to be fun. Everyone cleaning his gun". Apparently, however, none of the group took off for Lake City to join the fight.

Disappointment, dejection, exhaustion and intense physical exertion marred the enthusiasm of the intrepid prospectors in their quest for silver. Comments such as - "Begun commences to look worse for ware, same as the bal[ance] of us"; "Worked til dark – slept cold – no tent"; "Dejected. My hands getting pretty bad"; "All sore. Walked all night. Such is life in the R. Mountains, trying to make a stake to satisfy something" – give us insight into the intensity of effort and paucity of return involved in this rarely profitable enterprise.

The harshness of the weather and primitive environment could occasionally give rise to a bit of poetry, a welcome reassurance that the spirit within could not be so easily conquered. Judd gave voice to this on a Friday in May, writing, "Now if ever I must give my pen flight or rather turn it loose. Emblazon the sky with God's own fireworks and the resounding of a pork of artillery shocking even my conscience, but I will leave that to great writers. Simply saying, a Devil of a Storm".

Judd's happiness, if ever it existed, was frequently focused on his letters from Ada Clark. All that came to a crashing halt on May 23rd when he received the first of seemingly inevitable "dear John" letters. Telling him of her pleasure in spending a full day with a gentleman caller, she proceeded to say that she would not trouble him any longer. Broken hearted, Judd penned to himself, "Enjoy till I return. Scant pleasures for long woes are to succeed." As it turned out, the woes were to be his. His terse entry for July 11th was, "A.E. Clark married today. We blame inconsistency in women but only when we are the victim. We find it charming if we are the object." He never mentioned Ada again in his diary, and he never married.

Garwood Hall Judd's life changed forever on Wednesday, May 26, 1880 when, at one o'clock in the afternoon he arrived in Gothic City. He reports that at 2:30, "I paid Three Hundred for a lot of 25 feet on Main Street, a small building 25 by 14. Don't know if it will pay out, but I knew if I bought something would stop running round. My falt." The purchase thrilled him, as did sleeping with a roof over his head. The next morning he recorded, "Whoopee. Slept in my own house last night and thanks be to the 'noble' three hundred I slept warm." Buyer's remorse was not entirely absent from his glee, as he noted later that he thought he had been too quick. This first purchase, however, set him on the path of entrepreneurship and he continued to purchase real estate, both empty lots and some with buildings. He did not hesitate to relocate those buildings if he found their positioning not to his liking, moving some as much as thirty feet back from the street and connecting together buildings to make them larger. He opened a saloon in June of 1880, recording that he sold his first drink for 25 cents to Mike McGee on June 26th, and offers the names of some of his businesses as a "Tiger Club Room", a "Billiard Hall" and an "Assignation Room". "Think I am doing well for E.D.'s son," he bragged on August 26th.[6]

Judd appeared to take seriously his responsibilities as a landowner and businessman. He served in various town offices, including those of town trustee and police magistrate, to which office he was elected in 1888.[7]

Sadly, Judd's loss of Ada E. Clark to another man took the steam out of the keeping of his diary. After the 11th of July, there were no more than half a dozen entries. How his businesses fared, what came of his friends and travel buddies, his life from that point on can be charted intermittently through the 1880s from items in the Gothic newspapers, as long as they were around. He ran for public office, served on the town council and added to his personal real estate holdings. His saloon was quite popular, as was his house of "assignations". John C. Johnson met him in Gothic toward the end of Judd's life, as Johnson was establishing the Rocky Mountain Biological Laboratory in 1928.

Dr. John C. Johnson's personal recollections of Judd were blunt and sometimes amusing. He said he was pretty well considered "unforgettable". His "manner of speech was almost elegant or at least outstanding, and yet he never offended the Western Man's ears by being too nice, too Eastern, and thus too 'sissy', since he interspersed some mild cuss words from time to time."[8]

As people migrated away from Gothic, their mines and other properties were more often than not left unattended and unsold; who would purchase real estate in a town that was dying, or even already dead? So Judd found another outlet for his skills; he offered himself as paid caretaker to absentee owners, to which end he was successful. He was paid $120 per year to look after the affairs of one estate, and $25 a month for a while by another individual. Eventually, he served as an early version of real estate broker, handling sales, purchases and other occasionally questionable transactions.[9]

A somewhat colorful remnant of a bygone era, Judd was unique in having summered in Gothic for fifty years when he died.[10] He had become a living "curiosity piece" and by the 1920s people were seeking him out. "The old man loved people and entertained them with stories of the early mining days at Gothic." His cabin was a virtual museum, a testament to the history of the town. We are even told that in 1928 the Fox Film Company made a two-reel documentary movie about him called, *The Man Who Stayed*, which ostensibly "enjoyed wide circulation".[11]

A fixture in time and place such as Judd cannot be allowed to disappear into the mists of history, or so believed the good people of the Rocky Mountain Biological Laboratory. Shortly after his death, they placed a bench made of concrete and stone overlooking the lovely waterfalls atop Copper Creek. Those falls are today known as "Judd Falls". The bench was repaired and relocated to a safer distance from the ledge in the 1970s, and the stone seat was replaced with one of marble that had been brought over from Marble, Colorado by Dr. Kenneth Ogg.[12] It bears the inscription:

To the Wayfarer, Friend or Stranger,
This Seat is Offered in the Name of
"The Man Who Stayed"
Garwood Hall Judd
March 14, 1852 – May 15, 1930

The "Spirit" of Judd

The bench that sits atop the path to Judd falls has certainly provided respite for many a weary hiker over the years, but at least one long-time RMBL associate wonders if Judd occasionally visits it himself. Susan Allen (Lohr) was Director of the Rocky Mountain Biological Laboratory for ten years in the 1980s. For several of those years, she lived year round in Gothic. Each fall, she would find herself completely alone there for two weeks. The summer researchers had departed, the winter caretakers had not yet arrived and billy barr, the modern 'man who stays' and who built a permanent home just outside Gothic, was away visiting family or friends. She shares this tale.

In the late fall of either 1987 or 1988, during her weeks alone in Gothic, Susan was heading home to her cabin from her office after a long day of grant-writing. The town was deserted and very quiet. Ahead of her on the path she suddenly spotted an old man. He was sporting "a long coat – I can still see him – and boots, carrying a bucket and he had a slouchy, droopy old felt hat on. He looked a little shaggy." She called to him, "hey, hey, what are you doing here?" When he failed to turn around or reply, she followed him up the hill toward her cabin. Without so much as hesitating, "he walked right through the wall of Murray cabin!"

Susan, a scientist and skeptic, was rather jarred by this experience. It is not every day that one sees a ghost. She saw him only from the back but he certainly resembled the man identified as Judd in many old RMBL photographs. Although no one else has come forward with a "sighting", Susan is not alone in believing that the ghost of Garwood Hall Judd lives on in Gothic – even if it just through his fifty years of dedication to the place.

So, here exists another legend of the West – a man whose life is shielded by the passage of time and dearth of written materials recording minute details of who he was, why he was where he was, and what exactly he did there. Why he should be of interest to us today, other than as a curiosity, is deeply imbedded in the history of RMBL.

Beginnings

A drive to Gothic on the first day of autumn can be an ethereal experience. Traversing the three short miles from the end of the paved road in Mt. Crested Butte, the images along the way change with each turn in the road and turn of the light. At first the dramatic vista overlooking the twisting, winding East River in the valley below takes one's breath away. Upon entering the canopy of trees that covers the road for several miles, the sunlight reflecting upon and through the quaking aspen trees threatens to divert one's attention from the way ahead. It would not be difficult to find a wheel spinning off the side of the road were heed not paid. Bicycles compete with automobiles for a cherished drive through the golden tunnel. If the time of day is right, a deer might pop out in front or be caught grazing seemingly inattentively in the open space. Diverse creatures tempt death as they scurry to and fro, frantically preparing themselves for winter, collecting seeds and nuts and whatever it takes to sustain them through the frigid months of darkness and snow. But the aspen are oblivious to life below. They reach for the sky, seeking and securing the glorious sunlight of early fall, showing off against the blue of the Colorado Rockies ceiling. Gold is a color ascribed to these dying leaves, but there is so much more than just "gold". Shades of red, orange and yellow make themselves known as the breezes ruffle the leaves. But gold is what we take away – pure, glorious, brightly shining gold. One wonders if it can be seen from space, so magnificent it is.

The Surroundings

At the start of the twenty-first century, most people do not aim their vehicles to Gothic as an end point. The goal of the drive is more than likely Emerald Lake, which sits above the town site, Judd Falls, a gentle hike from the parking area just above the town, or perhaps a trailhead farther along the dirt road. Copper Lake beckons, as do passes known as East Maroon, West Maroon, Frigid Air, Conundrum – some leading to Aspen for the hearty of foot, others simply to high mountain fields of brightly colored wildflowers or hot springs hidden high and deep in the mountains. Some come to drive through, others to hike or bike, still others to fish. There is a tempting and dangerous drive past the Devil's Punchbowl to Marble and still more waterfalls along the way, as well as a relatively benign loop over Schofield Pass, through Paradise Divide, back to Crested Butte. Whatever the draw, each year sees more and more people passing through Gothic and, perhaps, pausing for a glimpse of the world renowned field research station, the Rocky Mountain Biological Laboratory, Gothic's sole proprietor and resident.

Referred to locally as RMBL, pronounced "rumble", the Rocky Mountain Biological Laboratory was started in 1928 by Dr. John C. Johnson, one of the first professors at what for years was known as Western State College in Gunnison, Colorado. The story of the field station in its early years is very much the story of John Johnson, his family, and a few loyal colleagues. RMBL is one of only a few major biological field stations in the country to be totally independent, unaffiliated with any college, university or other institution. It has grown to international prominence, respected for the excellence of the biological research conducted there by scientists from many parts of the globe, and for its role in educating generations of scientists. It annually hosts scores of scientists and students who produce dozens of pieces of research for publication every year, and who have made some very interesting, valuable and significant discoveries.

Begin at the beginning…

Gunnison, Colorado's intellectual tradition may be said to have begun in 1901 when, after more than fifteen years of wrangling, the Colorado legislature passed and Governor James B. Orman signed the bill establishing what has evolved into Western State Colorado University. The institution was initially chartered as the Colorado State Normal School and tasked with offering a course of study in teacher training. It took another ten years to acquire the land, design and complete the first building and equip it with furnishings, books and other necessities. Opening day was September 13, 1911.

Listed among the founding faculty of the school was one John C. Johnson, who held a Bachelor of Arts degree in biology and received an initial salary of $1,000 per year.[1] The first class graduated in 1912. It included five women, one of whom was Vera Adams.[2]

Despite years of struggle to survive, caused mainly by lack of support from the legislature, the Normal School thrived. A mixture of dedication by the faculty and administrators, enthusiastic support from the community and a growing reputation for academic excellence helped to pilot the young school forward and keep its doors open. John Johnson taught at the college until 1928, with time off to return to his own studies and earn a PhD in zoology. He held the office of Dean of the College and Professor of Biology from 1919 to 1928, and was Vice President during his final year, by which time the school had been renamed Western State College (Western).

Writing of himself in the third person, an habitual trait, Dr. Johnson recalls his arrival in Gunnison. "That morning, September 1, 1911, when the little Rio Grande narrow-gauge from Salida [Colorado] crested the continental divide at Marshall Pass, young John C. Johnson entered Colorado's western slope for the first time – a whole new world to him. In his initial academic position, he was joining others coming to help Gunnison's new Colorado State Normal School get underway. Born of Swedish immigrants in a sod house far out on the state's eastern plains only twenty years earlier, he was used to hard work. He welcomed the challenge."[3] Johnson was understandably proud of securing this good teaching position only "six months before I became of voting age in Colorado!"[4]

Acceptance and matriculation at the Colorado State Normal School were contingent upon completion of "the eighth grade of a grammar school, or its equivalent". Interestingly, applicants were required to provide evidence of "high moral character" as well as "freedom from any contagious disease". High school graduates were permitted to enter the advanced program and might complete all degree requirements within two years.[5] Qualifying for the first graduating class were five ambitious young women, one of whom was Vera Adams, to whom John Johnson would be married in 1915.

During his tenure at Western, Dr. Johnson is said to have been popular with faculty, students and townspeople alike. In 1923, when the school was renamed, he conceived the idea of building a huge "W" on the side of Tenderfoot Mountain, which was owned by the college.[6] Along with four students, he climbed the side of the mountain and outlined the letter with old sheets and trees. It took several trips to get it just right, but on May 2, 1923, Western President Samuel Quigley declared a holiday so that students and faculty could haul rocks and outline the letter permanently. Dr. Johnson recalls that, for "several years thereafter, May 2nd was the annual college holiday for whitewashing the letter. There was no road to the top of Tenderfoot Mountain in those days, so the upperclassmen decreed that the male students of the freshman class should carry the ton of lime up the mountainside… The 110 lb sacks of lime proved to be a real burden to the freshmen… [but] freshmen weren't allowed to be sissies then."[7]

In 1914, Dr. Johnson initiated one of the first scientific research projects to be undertaken at the college, an effort to determine how to control the insidious mosquito population in Gunnison County. He is quoted in the June 24, 1914 edition of the *Gunnison News-Champion* as saying, "There is only one thing that keeps Gunnison from being an absolutely ideal place in which to live, and to which we can unhesitatingly invite students and tourists. This one thing is the mosquito."[8] Johnson set forth some real, practical methods of dealing with the pests, including the annual draining of ditches, cleaning up of breeding places and using kerosene extensively, but the town and surrounding areas soon became disinterested in repeating the actions regularly and they were discontinued.[9]

In November of 1924, Western State College was awarded a grant of land located in nearby Taylor Canyon. These 160 acres from public lands were granted by the United States Congress through the efforts of Congressman Edward T. Taylor, and were to be used for biological study. It was Dr. Johnson who promoted the idea of a biological field station and he mentored the project from then until 1928, when he left the college. That plot of land would eventually resort to the public domain, but the idea of a biological field station was mobile and remained with Dr. Johnson.

Early in 1928 the news burst in Gunnison "like a bombshell," that the State Board of Normal Trustees, at its meeting in Greeley, had "summarily discharged Dr. Johnson C. Johnson, dean of the college since 1919, and head of the department of biology since the founding of the institution in 1911."[10] Charges had been filed by Dr. Richard Aspinall, the new president of Western, alleging "insubordination, lack of cooperation, attempting to form a cabal of teachers and pupils against his authority, the use of profane and bellicose language to him personally, and other matters..."[11] Dr. Johnson vehemently denied all the charges, retained an attorney, and challenged the dismissal.[12]

It would take a lawsuit to clear Dr. Johnson's name and eventually restore more than $1,400 in back pay.

It seems that Johnson, along with college President Samuel Quigley, fell victim to political turmoil occurring throughout the state that was centered on the rise of the influence of the Ku Klux Klan within legislative and other government offices. According to Johnson's memoir, Quigley was fired because "he would not join the Klan and do its bidding. Neither would I, the Dean of the College, join the Klan. Members of the Klan, which included the local Gunnison member of the Board of Trustees of the State Colleges of Colorado, wanted to remove or demote the members of the faculty who were Roman Catholic. Actually, there were very few members of the Klan in Gunnison and Gunnison County. In order to make a big showing when they paraded in Gunnison in their white sheets they had to import Klan members from elsewhere."[13]

Johnson also was unable, or unwilling, to build a good working relationship with the new college president, Richard Aspinall, who had replaced Quigley, a man highly regarded by Johnson.

Unwilling to join the Klan or turn against people he respected, Dr. Jonson found himself sacked from the college as Dean and as a tenured faculty member on what he considered to be trumped up charges. Western's new administration may have been within its legal rights to terminate Johnson as Dean, with or without cause, but as a tenured faculty member he could not be fired without a valid reason. This initial dispute was settled early in 1928; Dr. Johnson was reinstated as Dean of the College and head of the Biology Department on March 21, 1928. He was, however, placed on paid leave of absence until August 31, 1928, at which time his resignation was to take effect.[14]

Dr. Johnson was incensed about the turn of events and vocal about the causes and the people responsible. In an undated and apparently unpublished letter, he took issue with an article published in the college newspaper, *Top O' the World,* in which the president discussed the settlement of the dispute. Johnson wrote, "The most important item of all was noticeably absent, namely, that the president of the college 'voluntarily' withdrew all charges and complaints that were manufactured against me and the Board of Trustees, therefore rescinded all its actions. Obviously no one withdraws charges and complaints that could be proved in court."[15]

Years later, in a long letter to *Top of the World,* Johnson took great pleasure in the fact that one of the new buildings at Western was named Quigley Hall, something he considered to be a long overdue honor for a man he greatly admired.[16] He also noted that Aspinall lasted only three years as president, that the size of the student body declined by more than 20% during his tenure and that "about fourteen of the best faculty members left the college during his [Aspinall's] term of office, including the writer."[17]

Financial difficulties plagued many colleges in the post-World War I years, and politics played a big role in which ones were to receive state funding in Colorado. Whatever the reasons behind it, Dr. John C. Johnson suddenly found himself without a job in early 1928. But, Western State College's loss was to become the gain of the small mining ghost town of Gothic, Colorado, future home of the Rocky Mountain Biological Laboratory.

An aside…

Harkening back to 1924, the 160 acre plot of land granted to Western State College by Congress had been designated as a site for biological field studies. As Professor of Biology at the college, Dr. Johnson was the driving force behind the creation of this field station, which was named the Rocky Mountain Biological Station (RMBS). It was close to Gunnison, only about twelve miles from the college, and required very little in the way of construction. One small building served as a work base there and it operated until 1927. Johnson had reached out to notable biologists to work at the site, receiving a positive response. Included among those who were interested were three from the University of Oklahoma – Dr. Aute Richards, Dr. Mildred Richards and Dr. A.O. Weese – as well as Dr. Henry B. Ward of the University of Illinois and A.E. Beardsley of Cornell University. "RMBS thrived."[18] And, though they didn't know it at the time, the founders of the future field station in Gothic were among this comradely group.

Meanwhile…

Back in 1919, Johnson had had his first glimpse of the small, ghost mining town of Gothic, Colorado. In June of that year he had received his Ph.D. in zoology from the University of California, Berkeley. He was, he said, "pretty well worn down physically and perhaps mentally… I promised my family that I would rest all summer…" Most of this "resting" was to be in Gunnison, Colorado, where he arrived on June 4, 1919. Resting for Johnson consisted of visiting his many friends, hiking and fishing. But, "after thus resting for about two weeks, I soon learned that, for a young vigorous man at least, resting was harder than working."[19]

Ever curious about the surrounding area and having heard about the ghost town of Gothic, he eagerly accepted an invitation from a friend on the Fourth of July to "take an auto ride" to Gothic. As they motored closer to Crested Butte, he was struck by the increasing beauty of the surrounding landscape. Yet further along, on the eight-mile drive from Crested Butte to Gothic, he was still more impressed. Arriving in Gothic in the early afternoon, "…the shadows were just right on the east side of Gothic Mountain, showing up the many and enormous cathedral spires and abutments. Several very large snow patches, one over a thousand feet long, were also very beautiful on the mountain, as were the abundant columbines, legumes, fireweed, shooting stars, gentians and numerous species of the *compositae* family of daisies, asters and their kin. Surely, I said to myself, this is a paradise for the biologist who wants to study the hundreds of species of plants and animals in a very beautiful and primitive area, almost exactly as the Indians had left it only forty years previously."[20]

He had no way of knowing then that Western State College would ask him, shortly thereafter, to return to the college as its first Dean and as Professor of Biology, or that only nine years later he would be founding and organizing the Rocky Mountain Biological Laboratory in that beautiful town site of Gothic.

Back to 1928…

After the difficulties associated with his departure from Western, Dr. Johnson accepted a position as Head of the Science Division at what is now West Chester University of Pennsylvania. Part of his agreement with this institution was that he would be free to return to the Gunnison country each summer "to be the Director of a Biology Field Station that I expected to organize in the high Colorado Rockies."[21]

Dr. Johnson had also harbored hopes of continuing to use the facilities of the RMBS, the field station he had founded in Taylor Canyon, but that was not to be. The new administration at Western would not grant Johnson, no longer an employee, access to the facility and claimed for several years that other biologists on its staff were planning to work there. After only a few years, however, the college decided to discontinue its support for the place. It was abandoned and the government grant was soon rescinded by Congress.[22]

Thus ended John C. Johnson's relationship with Western State College…

Well, yes – and no. The truth is that disagreements and litigation between Johnson and Western continued for several years. The college considered the name of what would become Johnson's newly created field station in Gothic to be too similar to that of the old one under its dominion – it was, in fact, exactly the same. Western had also stopped paying his salary. Johnson's attorney, Clifford Stone, filed a lawsuit on Johnson's behalf to recover his lost wages.[23] In turn, Western sued the new Rocky Mountain Biological Station for using the same name as the Station in Taylor Canyon associated with the college.

At least one colleague in Pennsylvania urged Johnson to drop his claim against Western for past wages and devote all of his energies to his new work.[24] Johnson's attorney, Clifford Stone, counseled that the costs of litigation weren't worth the sums Johnson had coming. For his part, Johnson was adamant. "I will simply not be robbed… in addition to being lied about and stepped upon – nor would you or any other self-respecting man. After all, a man who doesn't respect himself and trust his own judgment after weighing all things carefully is no man at all!"[25] At one point

Aspinall called for scientists from other colleges who had been affiliated with the RMBS project to turn away from Johnson; they refused. And, in the midst of this protracted period of conflict between Johnson and the college, Johnson continued with his plan to create a new, independent field station.

A New Beginning…

Johnson's years of teaching, study and involvement in the Gunnison community, coupled with his enchantment with the ethereal spot called Gothic, convinced him in his own mind that there was, indeed, a need for a biological field station in the Gunnison country, that it should be in Gothic and that he should be the one to found it.

That summer of 1928, his transitional year from Colorado to Pennsylvania, was also the formative year of what was to become known as The Rocky Mountain Biological Laboratory. Perhaps it was the less-than-happy parting he had experienced with Western State College, perhaps it was a dream, perhaps a conceit, but he was convinced that the field station should be totally independent of any college, university or other institution. Toward that end, he was determined to "be responsible for all finances involved, to risk what money I had and could borrow, to start this scientific adventure."[26]

Implementing great visions often accompanies willingness to make great sacrifice. Johnson's vision of a field station was one he acknowledged to be "a great opportunity for testing academic training and managerial skill". In later years, he admitted that, "If I had known the difficulties ahead, I now wonder if I would have had the courage to try".[27]

Along with his own, there were four additional signatures on the Articles of Incorporation for a new field station, dated April 27, 1928. Those were that of his wife, Vera Adams Johnson, and three colleagues, all of whom had worked with him at RMBS. From the University of Oklahoma were Dr. Aute Richards and Dr. A.O. Weese; and from the University of Illinois, Dr. L.A. Adams, whom Dr. Johnson credited with inspiring his career in biology. Legal services were donated by Gunnison attorney Clifford Stone.[28] Moral support arrived in abundance from friends and colleagues around the county and throughout the country.

Based on a vision, a dream and with a very small budget, the Rocky Mountain Biological Laboratory, a non-profit educational corporation of college-university rank, was born![29]

The Johnson Family

For many years, it could be said that the Rocky Mountain Biological Laboratory, Dr. John C. Johnson and his family were interconnected to the point that one could not exist without the other. The concept of an independent biological field station in Gothic, Colorado was originated by Dr. Johnson. He was responsible for finding the place to build it and assembling the other individuals who would constitute the original board of directors. His first wife, Vera, and his second wife, Mildred, were by his side through all those formative years and beyond. His children were actively involved in the physical running of the place. Some of his progeny would continue in the scientific tradition and spend their lives supporting RMBL academically and administratively; others would forgo science yet also work in support of the institution in different ways. Perhaps most importantly, the entire cost of the enterprise was underwritten personally by Johnson for more than two decades. What his family ate, how they dressed, where they traveled were all dictated by the needs of that little spot high in the Rocky Mountains. In many ways, John C. Johnson **was** RMBL, and his family has taken, and continues to take, their responsibilities toward his legacy seriously.

John C. Johnson Sr. (1891-1973)

John Johnson was born in 1891 in a sod house in the eastern plains of Colorado and grew up in the nearby town of Sterling. The original family homestead was located in Leslie, a small town that no longer exists, located about fifteen miles north and four miles west of Yuma, Colorado. According to family records and remembrances, Dr. Johnson's Swedish grandfather, Peter Johnson, emigrated from Sweden to the United States with his large family in June 1870. Dr. Johnson's father, John Johnson, was the oldest of seven sons and one daughter and was twenty-two at the time. The family apparently had some money, as they were able to purchase ocean passage for ten to New York and a large piece of property in western Iowa, near Council Bluffs. Hearing of "free" land available for homesteading in Colorado, John Johnson decided to move his wife, Anna, and their growing family to settle on a homestead in eastern Colorado, arriving in March 1888. There, in addition to working their land, the couple opened a store and set about raising their seven children, five girls and two boys. John was the second youngest of the lot and the first boy.[1]

Young John Johnson seemed to have enjoyed his early years in rural Colorado, claiming to have been "a happy youngster. The fact that I was the first boy in the family… gave me special privileges. All my sisters catered to me…" As an adult, he relived those years in memoirs written for his children and grandchildren. He had a strong sense of family and personal history and did not want it to be forgotten with passing time. He was aware of his Swedish heritage and spoke the language fluently growing up, but wanted to fit in with other American children around him. His memoirs are erudite and often whimsical. In one, he recounts how he came to have his middle name, Christopher. It appears that when the family relocated to Sterling, his sisters noticed that other children had middle names; none of the Johnson family did. "The children of Sterling had such pretty middles names, it seemed to us, so the six Johnson children did their best to have pretty middle names", he wrote in one of his journal notes. Each of them chose carefully so that their names sounded pleasant to them. Young John was about to complete first grade and had just finished reading about Christopher Columbus and his "daring exploits", so he chose both names and signed his name "John Christopher Columbus Johnson whenever possible". Their parents indulged the children in this whimsy, but after several years did suggest that John drop one of the names; "…so with tears in my eyes I dropped Columbus". For the rest of his life, he would be John Christopher Johnson.

In his journals, Johnson shared his memories of things that made an impression on him as a child. In one, he writes of meeting Buffalo Bill Cody twice. He records that Buffalo Bill called him "Swedie" and "Tuffy" - "which were my nick names if my playmates didn't call me something worse. How proud I was to talk to Buffalo Bill". He remembers collecting buffalo chips for cooking and heating, which "in those days were about all the fuel available". His memories of "barefoot days on the homestead and in Sterling" are spiked with cautionary tales, such as how he would drive rattlesnakes down a hole with a stick and put his foot over the hole to keep the snake from biting him. "The danger here is not that the rattlesnake could coil and strike me while in the hole, but rather that it could easily have done so while I was chasing it with a stick in my hands." He writes fondly of his days playing baseball, a very popular game among the children of Sterling. "For several years I caught without a mask, but then I was hit too often in the face – actually three days in a row from the same bat wielded by the same person." His father proceeded to acquire a mask for him, but he had already earned the nickname Tuffy - and the reputation that "you can't kill a Swede".Other stories that he shared for posterity include such titles as "I Almost Drown – Twice", "Riding Mustangs", "Making Strings of Beads from the Centers of Cactus Flowers", "Deserted on the Plains", "Bronco Busting Days" and "Snakes Die at Sundown". "The very first dollar I ever owned came the last day that I was in the third grade in Sterling Public Schools," he wrote. He was an ardent student, according to his recollection of himself, and was quite relieved and happy to be passed on to the fourth grade. But even "more joyous" was winning a silver dollar for the "greatest improvement in writing of any pupil in [the] third grade during that year". In retrospect, this is quite understandable as he continued to be a prolific writer throughout his life.

The sod house that young John lived in on the homestead was built by his father and, though modest, was larger than average. According to John's journal, it had two rooms and "a sleeping attic large enough to hold one bed above one of the rooms". It measured about 24x30 feet, approximately twice the size of the average sod house and, he said, it was so "beautiful that people came from far and wide by wagon and horseback to see this wonderful bit of sod architecture on the flat, wide open plains of Colorado."

Times were difficult and farming was especially difficult. Dr. Johnson's memoir points to the fact that early pioneers were unaware of the variations in rainfall in the plains of eastern Colorado. One year might be "wet" and another "dry", so that farmers were at a loss as to what crops might be brought in successfully in any given year, or if there would be grass for the cattle to eat. The farms in and around the family homestead began to fail and the elder Johnson found himself in financial trouble as his neighbors and customers began to abandon the area – and abandon their debts to him, as well. When young John was about six – in 1897 – his father gave up the store and traded the homestead for two lots in Sterling, Colorado, where he relocated his family. A proud man, he took a job as a janitor in the local school. The family was able to squeak by with the support of milk provided by a cow they owned. One year, not long before holiday season, the cow died and times became even tougher. But lo and behold, on Christmas morning the family woke to a new cow in their barn. The mystery of where it came from lasted about forty years, when a neighbor from that time finally admitted to having given them the new cow. She said they would not admit to it at the time because they knew that the fiercely proud family would not have accepted the gift; they were afraid their friends would have starved.

When the children began to reach the age of needing higher education, John Johnson moved his family once more, this time to Greeley, Colorado. He wanted all of his children to attend college. Greeley was a town dissected by railroad tracks. One day, when young John was in high school, his father was walking along "the labyrinth of tracks south of the depot."[2] As he crossed in front of a stopped train he failed to either hear or see another train coming from the opposite direction. He was hit by the second train and killed. Young John had to add the job of partial provider for his family to that of high school student. He delivered ice and worked at other odd jobs to help support his mother and siblings.

In 1911, John Christopher completed his studies at Colorado State Teachers College, now the University of Northern Colorado, in Greeley. He considered himself fortunate to be among the founding faculty of the Colorado State Normal School, which would become Western State College (Western), and, later, Western State Colorado University. During his tenure there, he was able to take time to continue his studies and in 1919 completed his Ph.D. in biology, with an emphasis on zoology and parasitology, at the University of California, Berkeley. Returning to Gunnison and Western, he was made Dean of the College and Professor of Biology, positions he held until his very public dispute with the new Western administration in 1928 and his departure from the college that same year.

It was during the period of uncertainty after leaving Western that Johnson hatched the idea of creating an independent biological field station in Gothic, Colorado. Through sheer force of will, perseverance and endurance he embarked upon a tour of colleges and universities seeking scientific colleagues and academic supporters. Successful in this endeavor, he was also successful in finding a new job. That fall, he took up his post as Professor of Biology and Chairman of the Division of Science at Pennsylvania State College in West Chester, Pennsylvania.[3] He remained in that post for ten years until, in 1938, he was abruptly dismissed.

The ouster of Johnson and two of his colleagues ignited a firestorm of anger in the West Chester community. Students threatened to strike, newspaper columnists cried foul, meetings were called, petitions were circulated, committees were formed and answers were demanded. A clipping from an unidentified newspaper dated Saturday, April 30, 1938, said, "From where we sit, the whole thing appears to smack very much of politics and not very clean politics at that… Our guess is that the real reason that the management is so anxious to get certain instructors out is… because those who run things want to get some others in."[4] The American

Association of University Professors in Washington, D.C. opened an investigation into the firing of the professors. But summary dismissal from an academic position in a teachers college was legal in Pennsylvania at that time; tenure did not apply. Through it all, the college kept a tight lid on the reasons why the faculty members were targeted for dismissal.

Finally, on May 11, 1938 the Board of Trustees of the college officially fired Johnson and ordered him not to appear on the campus again. The charge was insubordination in six specific instances. Johnson fought a public battle in the local press, but his sacking stood. A feisty man with the contentious departure from Western and the dismissal from West Chester behind him, Johnson was still able to secure another college-level position. He completed his academic career in 1959 after almost twenty years at the Edinboro State Teachers College in Edinboro, Pennsylvania,[5] retiring from that position with appreciation and accolades from the college Board of Trustees.[6]

Dr. Johnson had moved his family to Pennsylvania in 1928, but he returned with some or all of them every summer thereafter to the mountains of Colorado, to the little town of Gothic and the Rocky Mountain Biological Laboratory he had birthed, and which he nurtured until his death in 1973. He remained actively involved as Director until 1958, at which time he agreed to retire from that position so that the organization could be piloted by a younger generation of scientists into an evolving future. He did, however, retain his position on the Board of Trustees for another decade. He died in Denver, Colorado on January 12, 1973.

John C. Johnson was married twice and had four children, two natural and two adopted. Sadly, he lost his first wife, Vera, to cancer very early in life, when their youngest child was only six. But his second wife, Mildred, was a loving partner and dedicated mother to his brood. A tall, proud, somewhat reserved man, he experienced the good fortune of having a devoted family, which grew to include six grandchildren and at least ten great grandchildren. Known to be somewhat reserved in the expression of his emotions, Johnson is said to have told his new daughter-in-law, Dorothy, at the time of her wedding to his son Chris, that he bestowed hugs on only two people - his wife and his daughter, Clea. So Dorothy had to settle for a handshake after taking her vows!

Dr. Johnson was not incapable of sentiment and emotion; a most touching letter to his son, whom he called John and who was on active duty in Europe during World War II, is evidence of this. Dated "Edinboro, Pa., Nov. 28, 1944, 8:30 P.M.", he wrote,

Dear Son John:

It's twenty years to the very hour since you were born. In fact it's almost to the minute, since you arrived about 5:45 p.m. Gunnison time (three hours later here Eastern war time). It doesn't seem possible that twenty years have passed – and my how many things have happened, both of extreme sorrow and of extreme joy, and all stages in between.

I just wanted to say that you have contributed a very great deal to our joy and happiness, and we are very thankful for you, and proud of you. We appreciate especially your genuine manliness, honesty and integrity. May it always be so, and I know it will.

You know that it is our fondest hope and prayer that you keep well and safe, and then be restored to your family and friends. I feel confident that you will be, since you have so much to give to the world. It will always be the aim of your parents to help you achieve all the worthwhile ideals and ambitions of yours, and in the end you will have lived a most useful and satisfying life.

Always with our very best wishes and much love from your mother and Your Dad.[7]

This letter not only gives one a glimpse into the affectionate and loving soul of a father, but also a hint of Johnson's sense of detail. He had a memory that just didn't quit and could spout dates, names, places and facts about almost anything that had occurred in his life. He also proved to be a good judge of character.

Beloved by his family, he was equally respected by his colleagues and friends in Gothic. The 1973 RMBL *Newsletter* written in "memory of Dr. J.C. Johnson, Sr., Founder of RMBL", states that throughout his life "he devoted his thought, energy, and academic income… to the development of our laboratory and the preservation of its precious habitat. Dr. Johnson served continuously from 1928 in the various roles as Director, Teacher, President, Trustee, and lately as Founder, an office for which he alone was eligible." It goes on to add, "The selflessness of his service to RMBL was obviously contagious as manifest by the magnitude of voluntary service contributed by his family and friends…. While many amenities have been added to RMBL since 1956 by NSF [National Science Foundation] Grants, the existence of this laboratory stands as a monument to the vision and faith of Dr. Johnson."[8]

The aforementioned voluntary service continues to be a calling among the members of the Johnson family, all of whom have enjoyed exposure of some kind to Gothic and RMBL. Johnson's personal selflessness and dedication to the cause inspired others to give of themselves to build and maintain his beloved Rocky Mountain Biological Laboratory. His memory continues to be greatly revered by those in Gothic and beyond who knew him, and those who knew of him.

Vera Adams Johnson (1893-1933)

Vera Adams' grandfather, a Civil War veteran on the Union side, arrived in Gunnison with his family by covered wagon in 1881. Her father, Clarence Devers Adams, worked for a season herding cattle, but then found employment with the Denver, South Park and Pacific Railroad, working his way into the positions of fireman and engineer, running his train through the extraordinary Alpine Tunnel, which connected South Park and the Gunnison Valley. In July, 1891 he married Ethel Miller, the daughter of a Gunnison pioneer family. They had two daughters, Vera and Alta. Vera was born on January 4, 1893, her sister a year later, in Como, Colorado, where the family resided for a number of years.[9]

According to family notes, Clarence Adams suffered a near fatal accident with the railroad, prompting him to find a safer job for a family man. At the end of the nineteenth century, Adams brought the family to Gunnison. There, he purchased a mortuary and furniture store from his brother-in-law, Sylvester J. Miller, and became a licensed undertaker and embalmer. The family settled into a spacious home and into life in Gunnison. Vera and her sister attended Gunnison public schools and Vera completed high school in 1910. In 1911 at the age of 18, she enrolled in the Colorado State Normal School. It was there that she met the young faculty member, John C. Johnson. After graduating in 1912, she went off to teach in Naturita and Montrose, Colorado for a year. She continued her education in Greeley, receiving a Bachelor of Arts degree from the State Teachers College in 1915. On July 23, 1915 Vera Adams and John Johnson were married in Gunnison.[10] The couple went on to adopt twin boys, Clarence and Lawrence, two years before Vera gave birth to the first of their two natural children, John C. Johnson, Jr., known as "Chris" (1924-2012). Clea Marie Johnson was born almost three years later.

The love affair and marriage between John Johnson and Vera Adams is significant because Vera was not only John's partner in establishing the Rocky Mountain Biological Laboratory, but the mother of what might be considered a RMBL dynasty. Their son, Chris, would continue the tradition of leadership of RMBL for decades after his father passed away, and their grandchildren and great-grandchildren still lend the family's support to the organization. Sadly, Vera died very young, at the age of forty, in 1933.

According to Chris Johnson, Vera Adams Johnson "should be regarded with her husband as Co-Founder of RMBL, for only with her constant dedicated support could the Lab have been established".[11] When the Articles of Incorporation for the new field station at Gothic were signed on April 27, 1928, Vera's was one of the five signatures on the document. With four young children by her feet Vera, no scientist, managed to provide her husband with invaluable support during those formative years of the Lab. "Vera's energy, courage, humor and calmness were boundless. She could multi-task before the word was even invented. While John worked on founding the Lab, Vera kept the home fires burning – literally."[12] Along with her mother and sister, she worked to sew homemade curtains for the cabins and dining hall, as well as comforters for the beds. She generously loaned her personal pots, pans, dishes, cutlery and other household necessities to the Laboratory and oversaw their transport over perilous dirt roads, thirty-five miles to the town of Gothic. She gracefully accepted her husband's decision to use their personal monies to underwrite the costs of the institution, including purchasing real properties. About their finances, she wrote at one point,

> This family is about down to bedrock as the teachers' checks haven't come in yet which were due the first, so we are about living on bread, water and toothpicks, and if the money doesn't come in today I am afraid the diet will be bread, water and toothpicks minus the bread. Here's where a well filled cellar would surely come in handy.[13]

After the family relocated to Pennsylvania in 1928, they managed each summer to pack the car and transport the six of them to Gothic, Colorado and reopen the site. A special treat for Vera would have been to move into the family's newly constructed log cabin, known as the Johnson House. The cabin would be ready to occupy in 1934 but, unfortunately, Vera did not live to enjoy it.

During the course of the summer of 1932 and into 1933, Vera suffered what she was assured by her doctors to be a difficult pregnancy. However, after a lengthy ten months produced no baby, surgery revealed instead a large ovarian tumor. The tumor was removed, but Vera was too ill with infection and cancer to survive. She passed away on April 6, 1933 and was buried later that month in the family plot near Gunnison. The obituary in the local newspaper eulogized her in glowing terms. She was, it said,

> …a splendid woman… A sweet school girl radiating joy and

> happiness in her contacts with those about her. A capable and sympathetic teacher, wielding influence for good on the pupils whose steps she guided. Finally, a devoted wife and mother, center of a family circle where her presence will ever be a benediction.[14]

Although Vera was not a scientist, she has been recognized by family and friends alike as a founding energy behind the Rocky Mountain Biological Laboratory. She was, "a winner… central to providing the ambiance that surrounded and included everyone who came and became a part of the fellowship of those biologists in those first, early years".[15]

Mildred Fischer Johnson (1890-1984)

Alone at the age of 42 with a challenging academic career, the Rocky Mountain Biological Laboratory to run and four young children to care for, John C. Johnson was fortunate to find in his second wife another partner in life.

Mildred Fischer came from a prominent family in Philadelphia. She had earned a Bachelor's degree from the University of Pennsylvania and a Master's degree from Columbia University. She went on to serve as Dean of Women at Mansfield State College and at Pennsylvania State College at West Chester, where she met John Johnson. They were married in June 1934, "in the first wedding ceremony ever held in Phillips Memorial Chapel, on the college campus".[16] They departed for Washington D.C. on their honeymoon and then extended their trip westward "in time for Dr. Johnson to take up his regular summer teaching work at his [*sic*] Rocky Mountain biological laboratory, near Crested Butte, Colorado".[17] According to Chris Johnson, she "adapted well to the relatively primitive life at Gothic".[18] Or, at least she did so for a while.

One of the biggest challenges Mildred faced was raising the four Johnson children, who were elementary school age at the time of her marriage. The twin boys, Clarence and Lawrence, were quite a handful after more than a year of living without a mother. In the words of their sister, Clea, they were "very mischievous". Clea was only six when her mother died, Chris eight, the twins ten. Mildred, not quite a year older than her new husband, had never been married. She was quite sensitive about their age difference, which was not common in the era. She was, however, uniquely qualified to step into the role of mother. When she was in her twenties, a young family member was killed in an automobile accident, leaving behind a young baby. Mildred generously and lovingly embraced the child and cared for it for about a year. She and Dr. Johnson consciously decided to not increase the size of their family, although Mildred would have liked a child of her own. In the opinion of her husband's descendants, she sacrificed her own interests for those of her husband and the family she had inherited.

Mildred may have adapted to Gothic initially, but it was not to last. After several summers it was determined that she was not a good candidate for Gothic life. She suffered from what appear to be symptoms of altitude sickness – violent headaches, high blood pressure, nausea – when she was there. For many years, Mildred alternated summers in Denver, West Chester or Edinboro, PA and took up weaving to pass her time. She did move to Denver upon Dr. Johnson's retirement, but it took years for her to acclimate to even that altitude, which is four thousand feet lower than Gothic.

The Johnson family members have their share of tales about Gothic and RMBL; Mildred was no exception. Sometime after her marriage to Dr. Johnson, Mildred was chairing a fundraiser in West Chester, Pennsylvania, where they lived. In that role, she came across an American woman who had been married to a Russian count. Her name was Countess Julia Dent Cantacuzene Spiranski-Grant. The Countess claimed that her grandfather was President Ulysses S. Grant. Having heard the tale of Grant visiting Gothic, Mildred asked her if she was the little girl who had accompanied him. Indeed, she was his first born grandchild and had fond memories of her grandfather; she had traveled with him on that journey. Mildred did not record what Julia's memories were of Gothic.[19]

As had Vera, she served her husband in a strong, supporting role at RMBL; she was devoted to him, his life and his children. Chris's wife, Dorothy describes Mildred as a "second mother" to her; Mildred embraced her upon her first visit to Gothic in 1949 and refused to allow the pregnant Dorothy to sweep floors or climb hills or do any work that would stress her. The two stayed close throughout Mildred's life, even traveling abroad to Mexico together after Dr. Johnson died. Despite her problems with altitude, Mildred outlived her husband by a dozen years and died in Boulder, Colorado at the venerable age of ninety-three.

Clea Marie Johnson Greenawalt (b.1927)

Clea likes to tell people that she visited Gothic before she was born. Her mother traveled to Gothic when she was pregnant with Clea, who was born in May 1927. Although her parents left Gunnison in 1928 and she, therefore, doesn't recall anything about her first summers in Gothic, she does remember driving out there from Pennsylvania, all six family members in the 1927 Pierce Arrow, "every summer at the speed of 35 mph".

Early childhood memories can be vague and incomplete, but Clea does clearly remember certain things about her youthful summers in Gothic. At the end of each summer, all the buildings had to be emptied of their contents and transported to Crested Butte for secure storage, as otherwise they would be fair game for rodents – and for thieves. When the Johnson House was built, an indoor bathroom with flush toilet and hot water was included. Unfortunately, this was enjoyed for only a few years before all of the fixtures were stolen. It proved too costly to replace and outhouses continue to be the norm at RMBL even today. According to Clea, her father knew who had stolen the items but decided not to pursue reclamation.

Summers in Gothic were great fun for Clea and the other children. The Johnson children were expected to work at the beginnings and ends of the summers, setting up and tearing down the facilities. In between, however, fun times ruled! She recalls that each child had a horse for the summer. It cost very little to lease an animal, and the children were expected to feed and care for them. In those early years, free range cattle would wander into Gothic at night. A cowboy was paid by someone to watch the cows and each morning the children would mount their horses and ride with the cowboy to chase the cows back to the mountains. The horses were their legs. "We didn't walk anywhere; we rode our horses to breakfast, lunch and supper!"

The Johnson House in Gothic was notable for the catwalk that spanned the sleeping lofts on either side of the house. The children used to revel in maneuvering across it, with just the living area below them; there was only a log hand railing (at an adult's waist height) to hold on to, and no safety net below to break a fall. The catwalk had a utilitarian purpose, too. Gothic was so remote in those days and the road so rarely used by automobiles, that cars were easy to spot. During those times when the Johnson parents would go to Crested Butte for groceries or supplies, the children would take turns perching on the catwalk, watching for dust or headlights, signs that the parents were on their way home and it was time to calm down, clean up and behave.

As a young woman, Clea cooked in the dining hall, using pressure cookers on a coal stove. Perishable items were stored in the nearby spring and Dr. Johnson made frequent trips to Crested Butte for supplies. Food was a necessity, not a luxury; canned goods and grains figured prominently on the menu. Clea clearly remembers the delightful dishes that she and others prepared: "For breakfast, there was either oatmeal – or oatmeal. Dinner and supper were the same way." Canned or rehydrated dried fruits soon became something to be avoided.

Clea grew up in Pennsylvania, where her parents worked at the college in West Chester and her father at Edinboro State Teachers College. She graduated from Edinboro with a degree in elementary education. This is where she met her husband, John W. "Jack" Greenawalt. Jack originally had no interest in science; he intended to major in coaching sports. At Edinboro, he took one class from Dr. Johnson and switched his major to biology. He went on to earn a Ph.D. in bacteriology, worked at Johns Hopkins University medical school doing research and teaching, and was head of the Bio-chemistry Department at the University of Tennessee until his death in 1977.

Clea continued to teach and raised her three children, Jeff, Karen and Jenny. Her last position was as Assistant Director of Graduate Admissions and Records at the University of Tennessee. Through the years, she and her family traveled to and worked in Sweden and Mexico, but made only occasional trips back to Gothic. However, she never lost her connection with Gunnison, Gothic and RMBL. After retirement in 1990, she settled in Gunnison; "This is home to me," she says. For years after that, until oxygen became an issue, she would spend one day a week at the little, historic store at RMBL. "I would talk to people – I didn't do any work!"

John C. "Chris" Johnson, Jr. (1924-2012)

The first natural born child of John and Vera Johnson, Chris arrived after nine years of marriage, two years after the adoption of his twin brothers, Lawrence and Clarence. If his father was the founding spark of RMBL, Chris was the continuing energy. Studying science as an undergraduate at Ohio State University, he followed up with earning an M.S. and Ph.D. in Ornithology from the University of Oklahoma, focusing specifically on parasites in birds. Another RMBL "bugologist"![20]

It seems that Chris' entire life was centered in and around the Rocky Mountain Biological Laboratory. With his siblings, every summer from the time he was four was spent roaming the hills, mountains, woods and streams of Gothic, Colorado. As a faculty member in the Department of Biology at Pittsburg State University in Kansas for over thirty years, he continued his annual summer tradition at RMBL, serving there in many positions. He was Director of RMBL from 1968-1977, on the Board of Trustees and, until his death, the organization's self-appointed archivist.
Not only did Chris grow up at RMBL, he was among the second generation of scientists who brought their own children to Gothic each summer, resulting in what has been repeatedly called by the scientists and

their families alike, "one big family". His children followed in his sister Clea's tradition by being in Gothic before they were born. His wife, Dorothy, visited Gothic for the first time in 1949, pregnant with their first child, Carol. Carol and her brothers, Doug and Dan, spent almost every summer of their childhood enjoying the great outdoors of the Rocky Mountains that are the backyard of the little cabins in Gothic.

Chris Johnson passed away on March 2, 2012 but left a legacy that will never be forgotten by the community that is called the Rocky Mountain Biological Laboratory. Indeed, RMBL may be said to exist as a community today because of Chris' gentle guidance and enduring optimism during troubled times. His daughter, Carol Johnson, profiled him during a well-attended memorial service on July 29, 2012, held in the dining hall at Gothic.

> Dad... had a long and happy life, a beautiful life. But I am thinking about the things that Dad did NOT do. He did not use profanity... He did not ever – and I mean EVER – make negative comments about anybody... He didn't tolerate gossip... He was a really good listener... He never questioned the life choices that my brothers and I made – he was always just there with his car keys and his check book... Dad was a guy who did his duty, and was never interested in taking the credit. He was actually very modest and unassuming; he didn't have a big ego.[21]

Carol recounts that when her father was Department Chair at Pittsburg State, where he taught for over thirty years, he had a quote in a picture frame above his desk; he had the same quote framed at home. The quote is by Ralph Waldo Emerson and says, "There is no limit to what can be accomplished if it does not matter who receives the credit". Underneath, Chris had appended, "or the blame".

After completing high school and prior to beginning college, Chris Johnson volunteered for service to his country during World War II, where he was wounded during the Battle of the Bulge. He was awarded the Combat Infantryman Badge and the Purple Heart for bravery. A modest man, his family describes his exploits during the war. Chris served in Intelligence and Reconnaissance (I&R) and worked behind German lines. Prior to the Battle of the Bulge, his team was stuck behind those enemy lines and had to work their way through them to get back to the Allied troops. His wife recounted that he and a young German gunner came across each other and were able to communicate in French. They agreed to not harm each other and the German told them where to find the landmines that had been planted in the area. Chris and his team were able to navigate to the safety of their own side, only to be reassigned to combat in the Battle of the Bulge, where Chris' arm was severely shattered by shrapnel. He was returned to the States, finally arriving in Cambridge, Ohio, where he completed his recovery relatively close to his home and family. During his recuperation, he applied to colleges and settled on Ohio State University.

Dorothy Greenlee was already in residence at Ohio State. Chris moved into a tiny apartment owned by the Presbyterian campus ministry, where he paid for his rent doing custodial work. Dorothy's dorm was right next door to the ministry, so it seemed natural that they should meet. They dated for about a year and half before getting married on July 3rd, 1948. According to Dorothy, Chris credited that German soldier who helped save his life for meeting her.

For the first year of their marriage, Chris and Dorothy moved to Puerto Rico where Dorothy had been recommended by the national Presbyterian mission to assist an aging Presbyterian minister. Chris was able to find work as a teacher at the Air Force base there, but after a year he determined that he wanted to continue his education. He had been a science major, but unsure of his direction. It was the experience of teaching "a class every hour, all day" in different scientific disciplines that decided him in the direction of a Master's and Ph.D. in biology. To begin his new direction in life, he spirited his pregnant young bride from balmy, tropical Puerto Rico directly to 9,500 feet up in the Rocky Mountains, where she tried desperately to keep warm in layers of clothing.

After a summer of taking classes at RMBL, Chris and Dorothy settled into graduate school at the University of Oklahoma, where Chris earned his M.S. in 1952 and his Ph.D. in 1957. He joined the biology faculty at Pittsburg State University in Pittsburg, Kansas in 1956 and remained there until 1987, when he retired as professor emeritus. Dorothy was delighted to remain in that one place for all those years; "I don't like to leave where I am!" she claims.

Affectionate stories and descriptions about Chris abound at RMBL. One of the most common is that of Chris, pipe in hand or mouth, silently listening to someone talk or contemplating the solution to a problem. The current Director of RMBL, Ian Billick, recalls his first summer in Gothic as a student in 1988. His outstanding memory is that of Chris Johnson, then the Director, emerging to help make sense out of the chaos that could be Gothic to the uninitiated.

> With a pipe in hand, a blue jeans jacket, a gentle smile and thoughtful manner, he directed me to some tasks, probably shelving

journals and sweeping the mud out of the library. These tasks were undoubtedly more a way to provide financial help to a student than to ensure the efficient operation of RMBL. And while my work in the library was a very small part of my summer, my personal growth and discovery was that much greater because it was grounded by the RMBL community, a community of which Chris and Dorothy Johnson formed the core.[22]

Chris was known not only for his gentleness and mannerly behavior, but also for being quite conservative in his personal views or, as his daughter Carol put it, "he had standards". He had a pretty rough time of it in the late 1960s and early 1970s, when young graduate students with more liberal views toward social issues began to arrive in Gothic. Chris was adamant that unmarried couples not cohabitate. He refused to assign cabins to any couple that could not produce proof of their marital status. Naturally, clever graduate students found ways around this; many simply slipped off into relatively private areas of the town – generally so private they were known to all. To be seen walking away from the library attic during the wee hours of the morning was a clue as to where someone had spent the night, with someone else. But perhaps the most creative subterfuge was when one couple posted on their cabin door a copy of the Colorado statute allowing people to "marry themselves". This, of course, became all the rage! Faced with the statement of this, Chris was forced to swallow his principles and give in to the younger, hipper crowd.

Chris has been accused of having some eccentricities. He never walked from his cabin to the dining hall or his office; he always drove. For one who has experienced Gothic and understands its small size, this is quite amazing. There are only very few "roads", but Chris made them work for him. Chris loved having music and musicians around, but he did not like to dance. He thought it a grand idea to have Friday night folk dances and polkas, but he himself could only waltz – and he did that reluctantly. It was simply important to him that the young people summering at Gothic not be enticed into Crested Butte for fun and drinking. Still, he did not allow alcohol at the parties and dances, with the possible exception for a musician or two from town who would normally imbibe while performing. He wanted everyone to feel comfortable and at home.

Chris had a sense of humor and enjoyed telling jokes – repeatedly. No need for a new joke when you could count on a new audience each summer! He has been described by the researchers and people in town who knew him for many years as parental, steady, measured and caring. He would often meet the young women when they arrived in town at the beginning of summer and let them know where they would be staying that year. Or, he would just be there to say hi, welcome "home", a fatherly figure they came to appreciate.

Chris Johnson inherited some of his father's attention to detail and equally deep pride in RMBL. For many years, there would be a Fourth of July fish fry in Gothic which townspeople, forest rangers and visitors alike were welcome to attend. Legend has it that Chris would round up "volunteers" from among the RMBL students to give tours of the historic property to visitors. However, he first would spend the entire morning educating those people about the history of the town and RMBL. They would have to learn in a few hours' time the names and histories of all of the cabins and other buildings, plus the names of the five founders of RMBL, the universities they were from and probably even the names of their spouses. Then, he would grill each student to ensure he or she had gotten it right. Only then would he allow his volunteers to be set loose on hapless visitors with probably more details than any of them thought they wanted to know, but with plenty of enthusiasm for their task.

Chris' reputation and influence was not just limited to Gothic or the Rocky Mountain Biological Laboratory. He continued his father's practice of building good relations with the people and businesses in the local communities. Mt. Crested Butte, as a town, was late in coming to those mountains, so business in general was conducted with the tradespeople of Crested Butte. The Johnson families did not seem to be able to conduct business without building a friendship. Those friendships outlasted the businesses. One local woman, Trudie Yaklich, whose family has deep roots in Crested Butte, recalled that when her mother died, "the Johnsons sent Doug [their son] to represent their family at her memorial service". She still remembers Chris and her father "communicating through the puffs of their pipes. They just stood next to each other and puffed on their pipes. That seemed to be all they needed."[23]

On the serious side, Chris is remembered for "his character as a leader" of the Rocky Mountain Biological Laboratory, deserving of "immense credit of various kinds, professional as well as personal." When he became Director, it was toward the end of what had been the most difficult period ever in the history of RMBL, an era of conflict based on factions that were created by the changing practice of science and the requirements of the institution. One long time researcher at Gothic said of him,

> If any one person can be said to have saved this institution from the consequences of that very disturbing time period, it was Chris Johnson. He did so with a number of character traits that have been

mentioned already: by being very steady; by being willing to talk to and, in particular, listen to anyone of any viewpoint; and to try to provide whatever reasoned commentary he could muster. That was one of the critical features of his leadership.[24]

Despite his reputation for conservatism, Chris was recognized as "a man who could observe the need for change, think it over and make a decision to move forward, even in the face of challenges".

Chris Johnson did not want to be Director of the Rocky Mountain Biological Laboratory. He understood, however, that forces greater than his personal will were driving the survival of the institution. He trusted in his father to pilot the ship, but also saw that change was necessary to adapt to changes occurring in the field of biology.

Chris was a biologist, a recognized professional in his field and in his home institution of Pittsburg State University in Kansas. He was also recognized as an administrator of the institution that was the Rocky Mountain Biological Laboratory. Chris, however, had a partner in life, his wife Dorothy, without whom his role at RMBL would have been circumscribed. When one listens to tales about Chris, they are rarely if ever about Chris – they are about "Chris and Dorothy".

Dorothy Greenlee Johnson (1926-2015)

Dorothy Johnson first arrived in Gothic, Colorado in the summer of 1949. She was brought there by her husband of one year, Chris, and was expecting their first child. Although it was summer and ostensibly warm, the weather in Gothic was about as opposite as could be to that of Puerto Rico, where the young couple had lived the past year. Days might be sunny, or they might be rainy, or windy, or even possibly snowy; nights, however, were invariably cold and the warmth of a fire was needed to keep one comfortable. Oxygen was scarce at 9,500 feet, and any small exertion could leave one gasping for breath. There was no indoor plumbing. Food was, at best, meant to sustain one, not provide pleasure. A less intrepid person might have been concerned about the effects of such an environment on a developing fetus; but no one ever accused Dorothy of being of being fainthearted or unadventurous. As a matter of fact, she was particularly suited for the environment that surrounded the Rocky Mountain Biological Laboratory.[25]

Dorothy was born and raised in eastern Ohio. Her parents were educated people; her mother, a Phi Beta Kappa designee, taught school and her father, a graduate of Muskingum College in Ohio[xxvi], served in various posts, including as county school superintendent, an appointed political office for which she said he was well suited. He held his post from the 1920s until he retired in 1956. Though they married late in life, they bore four children and managed to provide quite well for their family during the Depression years and beyond.

The little mining town of Neffs, in eastern Ohio, was home to the Greenlee family during the era known as the Depression. Dorothy's father was paid in "paper money" for the work he did – she says "it wasn't real money, just a promise to pay". They owned their house in town and were accordingly able to rent it and earn some money from that. Dorothy's mother had been given a plot of forty acres of land by her father, who wanted to keep his children close to him. The Greenlee family relocated to the summer cottage they had constructed on that land and eked out a relatively comfortable living during a time when many did not. The "cottage" was a four room house with a basement and "would-be bathroom". As the bathroom was not completed, an outhouse sufficed. The kitchen was equipped with a coal cooking stove, which also provided heat, and the tiny living room was outfitted with a small potbellied stove. Her parents slept in one of the two small bedrooms with their baby boy; she and her brother shared the other bedroom and her sister slept in the empty, "would-be" bathroom.

The house had been constructed as a summer cottage, so it was not well insulated against the cold of winter. Dorothy recalled waking in the morning after a blizzard, her blanket covered with a layer of snow that had been blown in through cracks in the ill-fitted windows. Her father would rise first in the mornings and start the stoves in the living room and kitchen; then her mother would start the day with a warm breakfast for them. She said, "We were quite spoiled", but went on to recall that the children had to go outside to the pump to get water – a challenge in mid-winter. And, of course, there was that outhouse…

The family raised cows and chickens and in summers tended a large garden. Her mother "canned, canned, canned". The children were paid a penny or two to pick strawberries in season, which her father then managed to sell. She learned to churn butter and considered that they had a pretty good life, "because we had so much good food to eat".

School was a one room schoolhouse with one teacher. It was about a mile walk each way to get there and back ("not 'uphill both ways' because it was so flat"). Mostly the children walked unescorted – life was pretty simple and safe in those days and in that place. Dorothy recalled one year there was a severe blizzard and they were snowbound for two weeks. "We had a lot of fun – I don't know that the grown-ups did!"

Dorothy credited those years of country living and hard work with her ability to settle so easily into life in Gothic. "Eventually, because of that experience and the fact that I lived under those conditions, when I arrived in Gothic in 1949, my background prepared me for living in cold, and heating with wood and coal, and having to go outside to get your water."

When she arrived in Gothic, she came under the protective supervision of her mother-in-law, Mildred, whom Dorothy acknowledged as a real blessing to her, and like a mother. The doctor in town who was consulted about Dorothy's pregnancy advised her that walking downhill was fine but to not walk uphill. Dorothy was quite perplexed by this; "I thought, however am I going to get back to where I started?" In any event, she was constrained in her activities, but put in charge of bedding for the students at Gothic.

> I was to pass out three wool blankets and if somebody looked really skinny and like he'd be really cold, I'd give him a fourth. But they had flannel cotton sheets and they got a pillow and pillowcase [which] they were supposed to bring back to me every week and get clean ones. We'd send these into town and a coal miner's wife would wash them in a Maytag washer with a wringer. Our clothes went in there, too. They'd come back neatly folded and ironed in a cardboard box. That's how we got along without any way to do the washing.

Dorothy and Chris met while studying at Ohio State University. When it became obvious they would have a future together, they took each other home to meet the parents. Her visit to the Johnson home was quite an education for her, in more ways than one. It began with confusion as to who was who when addressing her future father-in-law or her future husband. She knew her fiancé as Chris, but the family called him John or Johnny. His dad was John, the name used by Mildred, but some close friends called him Christy. It seemed that young John Christopher Johnson, Jr. was tired of being confused with his father, so he unilaterally decided to use his middle name when he left home. He continued to be John or Johnny to his family and anyone who knew him before the war; but he reinvented himself from his college years on as Chris, and "that was who he became".

During that same visit, Dorothy observed Chris' younger sister, Clea, dying some clothes in a pot of boiling water on the stove. The family was so focused on and dedicated to underwriting the costs of RMBL that they cut corners anywhere they could, including dying or altering their clothes to make them look "new". Dorothy was so impressed by this that when she returned to school she determined to dye a wool suit another color so she could have a fresh outfit to wear. Nobody told her that wool shrinks in boiling water; that was the end of that suit – and of Dorothy's efforts to recycle her clothes in such a way.

Dorothy came from a long line of Presbyterians, some ministers, some church founders. Devoted to her church, she responded to the mission call more than once. The summer before her senior year of college, she worked with a Presbyterian nurse in northern New Mexico. The nurse spoke no Spanish and the community was populated by Spanish descendants who spoke no English. Dorothy spoke some Spanish and was able to communicate with the families and help in the inoculation program for the children. During that summer, she met a minister from Puerto Rico. He recommended her to the national mission program to work as an assistant to an aging Presbyterian minister on that island. After she and Chris were married, they sailed off to Puerto Rico and there they spent their first full year together.

Dorothy served in many capacities at RMBL, first and foremost as a wife and mother of three young children. She later stepped in to serve as the unpaid supervisor of the RMBL dining hall, planning menus, ordering food and overseeing the kitchen staff. Somewhere along the way, she morphed into somewhat of a maternal figure for many of the young researchers in Gothic. She and Chris were a team; he told her no secrets, she told none to him. They were the repositories of trust for the many differing concerns of the many differing individuals who crossed their threshold. They held their tongues and kept the peace. Their door was always open, coffee and tea were always ready to be served. They continue to be revered by the community in its entirety.

Dorothy Johnson passed away peacefully, surrounded by her children, on February 20, 2015.

These backgrounds are mean to be just that. The numbers of stories in which these individuals are included are countless, and they also include many other individuals. Some of these tales will be recounted in the pages that follow.

Not to be Forgotten

It is true that the Johnson family was the heart and soul of the Rocky Mountain Biological Laboratory, but they alone were not the breath. A little money and a lot energy go only so far without the support of a committed and supportive band of colleagues and friends. There would be many along the path to the present, but in the beginning there were three. Aute Richards, A. O. Weese and L. A. Adams were co-signators to the articles of incorporation of the Laboratory (or the Station, as it was known in 1928). Their academic connections, enthusiasm and willingness to work without pay all contributed to the success of the venture.

Aute Richards (1885-1974)

Born in Tekamah, Nebraska, Aute Richards moved to Kansas early in life with his parents. His undergraduate education was at the University of Kansas, and he later earned an M.A. degree from the University of Wisconsin in 1909 and a Ph.D. from Princeton University in 1911. He taught at the University of Texas in Austin and at Wabash College before settling into the University of Oklahoma in 1920, where he remained until his retirement in 1950. During his years in Oklahoma, he also served as Director of the Oklahoma Biological Survey and the Oklahoma Museum of Zoology.

Richards' field of interest and expertise was zoology. At the University of Oklahoma he served as Professor of Zoology and then as Chair of the Department of Zoology, teaching classes in comparative embryology and cytology. His research at the Naples Biological Station in Italy and the Marine Biological Station at Woods Hole, Massachusetts resulted in his book, ***Outline of Comparative Embryology.***[1]

Richards, along with his wife Mildred (who held a Ph.D. in genetics from Columbia University), had been an active supporter of the biological station operated by Dr. John Johnson during his tenure at Western State College. When Johnson parted ways with WSC and decided to pursue his own ideal of a biological research station, he turned to Aute Richards for support. Richards was happy to oblige and served as the first President of RMBL and Trustee from 1928 to 1938.[2]

A. O. (Asa Orrin) Weese (1985-1955)

Weese was born and raised in Hutchinson, Minnesota, and earned his undergraduate degree from the University of Minnesota in 1909. After an eleven-year stint in the department of biology at the University of New Mexico, he returned to school and earned an M.A. and Ph.D. from the University of Illinois. In 1924, following two years of teaching at James Milliken University, he began his professorial career in the department of zoology at the University of Oklahoma, working with Aute Richards. He was interested in many areas of study, but particularly insects of the grasslands. His summer work revolved around several summer research laboratories, heading up marine ecology at the Friday Harbor Laboratory in Puget Sound in 1925 and 1929. He joined John Johnson and Aute Richards as an incorporating director of the Rocky Mountain Biological Laboratory, where he developed the curriculum in ecology, taught summer classes and served as trustee. In 1938, he succeeded Richards at president of RMBL.[3]

L. A. (Leverett Allen) Adams (1877-1976)

L.A. Adams was well known and revered as a zoologist, working as Professor of Zoology at the University of Illinois from 1921 to 1948. From 1939-1948, he also held the position of Curator of the Natural History Museum at the Smithsonian Institution. He was, according to Dr. Johnson, influential in Johnson's decision to become a biologist. He was a co-founder of the Rocky Mountain Biological Laboratory and served as a trustee from 1928-1932.[4]

A Vision Realized

The year was 1928. It was spring and Dr. John C. Johnson had found himself without a job. He had time on his hands before heading east to take up a new position in Pennsylvania, so he conceived a plan. The biological field station that he had helped to create at Western State College was a thing of the past. The college preferred not to pursue it and the federal government would eventually repossess the land. But the idea lived on in the mind and heart of the man who had birthed it. He knew of another place, perhaps a better place – more remote, little touched by the incursions of civilization, or what substituted for it in this western region of the time.

In March and April of 1928, John Johnson decided to embark on a journey by automobile of what would become a 2,400-mile pilgrimage. Before the trip was concluded, he visited colleges and universities in five states – Colorado, New Mexico, Oklahoma, Texas and Kansas. His purpose was to inform the instructional staffs of institutions of higher learning of his plan to build an independent biological field station "a non-profit, educational institution which was to be devoted to advanced study and research in Biology and related subjects", and to try to interest them in joining him in his venture. He spoke with instructors in departments of biology, zoology and botany, and did not discount recruiting from outside these places. "Of course, the Biology instructional staffs in all the universities and colleges of the entire United States and Canada, and their advanced Biology students, would also be welcome," he writes in his *Brief History of the Rocky Mountain Biological Laboratory*.[1] His hope was that scientists might spend time in Gothic working on their own research in the areas of "high-altitude flora and fauna", and even that eventually some "foreign biologists might want to come and do research on appropriate high-altitude problems of the Rockies".[2]

Dr. Johnson's travels in 1928 brought him into contact with faculty of many colleges and universities, including the University of Oklahoma and the University of Illinois, where his idea was taken seriously by several individuals. It is only natural that the plan might have been met with some skepticism, as it evolved from the thoughts of one lone individual who had been fired from his academic positions. Where was the money to come from? Was there to be institutional affiliation? Would it be accredited as a college in its own right? If so, who would staff it? Where would the students come from? What, exactly, could be studied in such a remote and unforgiving location? And again – how was it to be paid for?

Understanding that there was no money available for this start-up enterprise, three individuals signed on as original incorporators, in addition to Johnson and his wife, Vera. The Articles of Incorporation of the Rocky Mountain Biological Station[3] were created by a Gunnison attorney, Clifford H. Stone, who cheerfully donated his time. In addition to crafting the Articles of this not-for-profit educational organization, Stone procured the Charter from the Colorado Secretary of State. The organization's Charter granted it the right to confer academic degrees, an option that has never been exercised. The date of incorporation was April 27, 1928 and the five original incorporators were

John C. Johnson, Ph.D., Chair, Division of Science, Pennsylvania State College at West Chester[4], Founder, Director, Trustee

Vera Adams Johnson, B.A., Wife of John C. Johnson

Aute Richards, Ph.D., Chair and Professor, Department of Zoology, University of Oklahoma, First President, Trustee

Asa Orrin Weese, Ph.D., Professor of Zoology, University of Oklahoma, President and Trustee

Leverette Allen Adams, Ph.D., Professor of Zoology, University of Illinois, Trustee

Johnson reports in his *Brief History* that he agreed to be responsible for *"all finances* involved". He paid to provide the required facilities, furnish them and cover the stipends of the instructional staff. He also agreed to underwrite the printing and mailing of an annual bulletin announcing courses, and to subsidize "all other necessary expenses until such time as the 'other members' of this proposed field laboratory might want to assume the financial responsibilities themselves". These 'other members' were asked only to "risk their reputations in associating with the writer [Johnson], and occasionally to be on the instructional staff, if needed".[5] It would be twenty-six years before Johnson's personal investment was repaid in full.

Now that the Rocky Mountain Biological Station existed on paper, it was necessary to make it a reality. The first summer session was to open on June 25, 1928 and there was much to be done. A complement of instructors with Doctorates was procured; an information bulletin was prepared and sent to "every college and university Department of Biology in the United States and Canada, and to the high school Biology Departments in most of the larger cities". The Johnson women – "wife, wife's mother and wife's sister" – busied themselves with sewing curtains and drapes for the windows of the cabins and making quilts for the beds. Johnson's father-in-law coincidentally and conveniently owned a furniture store in Gunnison. He supplied, at cost, the furnishings needed to make the cabins useable and comfortable. "Without this great financial assist it might not have been possible for the writer [Johnson] to have financed the heavy expense of getting The Rocky Mountain Biological Laboratory started."[6]

Dr. Johnson gives loving and generous credit to his children, Clarence, Lawrence, John Jr. (Chris) and Clea Marie for all their efforts to assist him.

> [They] contributed greatly to the Laboratory throughout many years by spending hundreds of hours doing the chores of the institution, such as chopping and carrying wood and coal to the cabins and kitchen, carrying water to same (before there was plumbing of any kind), filling kerosene lamps, and carrying and hauling furniture and equipment in their donkey cart and wagon. Perhaps their most appreciated service was helping the Director open up the laboratories and cabins, before the session began, and closing up of same, before there was any money in the treasury to pay for doing these menial chores. This opening and closing required about three weeks each summer.[7]

Johnson considered June 25, 1928 to be an auspicious and monumental date. Not only was it registration day for the first summer session but, also, unexpectedly a stranger by the name of Truman Blancett arrived in Gothic. Blancett was there, on horseback, for what he considered to be a nostalgic visit. He had been an early prospector in the area, pre-dating the actual Silvanite strike by a year, according to him, and taking six hundred dollars' worth of silver out. But this time he was fifty years older and just a visitor. According to Chris Johnson, son of the founder, looking back from the late twentieth century, "…on that one noteworthy day in 1928 a link was formed between the earliest recorded mining in this area and the initial day of RMBL's operation. It was a good, perhaps even providential omen, as if Blancett had magically appeared to pronounce his approval for RMBL to take over the reins, then for many decades, to 'dig' for biological knowledge".[8]

The physical plant at Gothic consisted of the broken down old buildings that Johnson had begun to acquire from Garwood Judd, as well as some other buildings that he initially rented from Judd, who claimed to be the caretaker of them. Eventually, all these structures would be acquired for RMBL by Johnson and later by the institution itself.

From the very beginning, the people associated with RMBL devoted themselves to recycling. Not that it was known as such in those days, but it was a matter of material survival. To discard a perfectly good piece of wood or log or useful item of any kind was sacrilege. Buildings that were deemed unsafe were sacrificed to reinforce more sturdy ones. Benches, tables and any furniture made by hand were the products first and foremost of wood that had been scavenged from old buildings and the surrounding environment. This would become a beloved tradition at RMBL – using the past to build the future.

The old town of Gothic consisted of about seventy acres, divided into 412 lots and containing approximately twenty-two old buildings in various states of disrepair, many of which would eventually be torn down and "recycled", if only for firewood. The conveyance to RMBL by Gunnison County in 1930 of most of the lots in the town in return for a payment of $100[9], and the purchase of more land in the 1950s, would bring the entire campus to a size of approximately 250 acres by 1960.[10]

On July 14, 1928, Dr. John C. Johnson, who had been temporarily appointed President, called to order the first meeting of the Rocky Mountain Biological Laboratory. The minutes of the meeting were recorded by temporary Secretary, Dr. A.O. Weese. "After much discussion concerning the probably worth-whileness [*sic*] of the venture", officers were elected. Dr. A. Richards was chosen as President, Dr. L. A. Adams as Vice President, Dr. A. O. Weese as Treasurer and Attorney Clifford A. Stone as Secretary. Dr. Johnson was named Director of the Laboratory, a position that he would hold for the next thirty years, until 1958. Dr. Richards was to continue as President for ten years, followed by Dr. Weese, who served in that capacity for sixteen years.[11]

Dr. Johnson was understandably proud of the quality of instruction at the Rocky Mountain Biological Laboratory, as well as the instructors who provided it. He boasted that from the beginning, "*every member* of the instructional staff has held the PhD or ScD degree and has occupied a good position in some university or college. Each course has always been taught by a highly qualified person in that field".[12]

Finances continued to be a challenge at RMBL, but somehow Johnson always managed to fulfill his commitment and pay the bills. Membership dues were only two dollars, and the organization's membership rolls were small for many years. Most summers, only between three and four hundred dollars remained in the treasury and sometimes, according to financial records, as little as $30. These funds were dedicated to building new student cabins, repairing existing buildings, purchasing necessary equipment for the laboratories or generally paying for whatever was needed to keep the physical plant operational. A second-hand generator was finally purchased in 1950, from the combined savings of two summers' revenues. This 3,000-watt generator was gladly welcomed after twenty-two summers of using kerosene lamps for light. Faculty members in residence at the time banded together to dig holes for the posts and string electric wires on them to the cabins, dining hall and labs. "One was an excellent electrician; and another, a fine plumber." This generator served as the sole source of electric power for five years, until 1955, at which time "a commercial electric system was installed from grants received... from the National Science Foundation and from contributions by the Rural Electrification Association". These were among the first grants and contributions received by RMBL in its initial twenty-five years of existence.[13]

Indeed, for the initial two decades of the life of RMBL, no one received a salary for his or her efforts, be they teaching or administrative. In exchange for their contributions, they received partial board and a place to live. Couples and families were accommodated with private facilities, to the greatest extent possible, and individuals were assigned to share faculty cabins. Men and women were carefully segregated. Dr. Johnson liked to say that, "It can truly be said that the R.M.B.L. was made possible and existed almost solely financially from the salaries that the instructional staff and the director *did not get*." But Johnson believed that they received their remuneration in different ways.

> No doubt among the great attractions toward being on the instructional staff of the R.MB.L. are the wonderful scenery and climate of the Gothic area in summer. Most of the days are bright, sunny but cool, seldom getting above eighty degrees. Then, the chance to associate closely with other biologists – students,

> instructional and research personnel – from other universities and colleges in many states is truly an unusual opportunity. Friendships made here seem to be lasting.[14]

And this is a refrain that continues to echo today. After more than eight decades in continuous operation, the Rocky Mountain Biological Laboratory has become "home" for dozens of scientists and their families. Children grew up together and returned as scientists in their own right or found other ways to continue to be involved and supportive. Families reunite each year in research, instruction and tale-telling in the dining hall or around the fire at night. The bonds forged in Gothic transcend time, distance and place. A most frequently uttered statement a curious inquirer hears in response to asking what people find so special about RMBL is, "This place is home".

A Sense of Place

From the very first time that John C. Johnson set eyes upon Gothic, he knew he had discovered something special, something rare – a place of spectacular biological diversity, containing innumerable species inhabiting a wide range of ecological zones and with vegetation of all types – montane, subalpine and alpine. Aquatic and terrestrial wildlife abounded, including large and small mammals, birds, insects, salamanders and fish. A pristine environment consisting of lakes, streams, fields, rocks, mountains, valleys - all seemingly as unaffected by intrusive human presence as it might have been in "Indian" times – except, of course, for a few short years at the end of the nineteenth century.[1] What wonders might it all reveal!

Consciously aware that "the whole area around this deserted mining camp was literally a paradise for both botanists and zoologists," Dr. Johnson began to lead groups of interested students to Gothic each summer after his initial visit in July of 1919.[2] A journal entry describes these weekends in his words.

"The group usually arrived shortly after noon on Friday, in sufficient time to make camp and get in a good field trip before supper. Since there was a flat grassy spot near Copper Creek back of the 1880's livery stables, we quickly erected our small tents, spread out our bedrolls or sleeping bags, and prepared the ample evening meal.

"After washing our dishes in the clear cold water of Copper Creek, we gathered around the camp fire for talk about what we had collected in the afternoon, for a bull session about most anything that seemed important, and for singing the old familiar western camp songs. Gothic would really ring with many barbershop chords."

"One time in about July 1925, it poured rain during the night and altho [sic] we came prepared for rain, and cut small drainage ditches on all sides of the tents, we were not prepared for that much rain. Noticing this, Garwood H. Judd, who called himself 'the man who stayed', said to me 'you should own this building so that you could at least have a place of shelter from a heavy rain, or from frosty mornings and cold nights.' Upon learning that he owned the building, I said 'How much?' 'One hundred and fifty dollars' was the quick reply. 'Sold' was the equally quick answer. Soon ownership papers were made out in my name.[3]"

The building would come to be called "Swallows' Nest", named for the cliff swallows that nested in the eaves over the front door. This two-room cabin had originally been owned and occupied by former Gothic Attorney McDonald. He had utilized the large front room as his law office, the back room as his living room. It is supposed that an attached lean-towas his bedroom.[4]

Swallows' Nest was Johnson's first real estate purchase in Gothic, but not the last. It was quite amazing how Mr. Garwood Hall Judd would offer up land and buildings as they seemed to be needed.

John Johnson was not a wealthy man, but he was determined that he should have a stake in the community known as Gothic. Whether or not the idea to preserve it as a biological field station was present in his mind in 1925, when he purchased his first building there, cannot be fully known. But we do know that by 1928, he had determined that this place should be saved for posterity and that he would be the one to do it, even at his own expense.

A year or so after purchasing the Swallows' Nest, Johnson ran into Judd on the road outside that building. He records that meeting in his journal, relating that Judd pointed out the old Gothic Hotel, also known as the Grant Hotel.

"You ought to own this hotel building; you could fix it up and use it for sleeping quarters for your gang of bugologists. You're getting too big for the house I sold you."

"I remarked, 'Yes, it would be a fine improvement' even though the roof leaked in a number of places, the doors sagged all over the building, and almost every pane had been broken out by vandals and just mischief makers who liked to hear the sound of shattering glass. Mr. Judd remarked, 'I will get it for you'. I said, 'that would be fine', thinking that he was just joking with me."

"A few weeks later we met very close to the same place on this same little roadway, and Mr. Judd remarked, 'Here is the deed'. I said, 'What deed'? His almost hurt reply was, 'for the hotel building of course'. I remarked approximately, 'Well, how did you get the deed"?

"Mr. Judd replied, 'I know the man who owned the building, and I told him to send me (Judd) the deed for the place since I know a man who really needs it for quarters to bed down his growing squad of college bugologists. So, of course he sent me the deed for the building. I (Judd) explained that he (the owner) had no further need for the building since it had not been used for many years and had gone almost to rack and ruin."

"I looked at the deed and sure enough it was made out to John C. Johnson

signed by Garwood H. Judd. I remarked, 'You surely didn't get the building for nothing'. 'Oh, yes I did', was Judd's reply. After getting over my amazement at this most unusual business transaction, I remarked, 'but the making and recording of the deed in the Gunnison County Court House must have cost something'. Judd finally allowed that it had cost him one dollar. 'What about the cost of the deed from the owner to you?' I remarked. He allowed that had cost him another dollar. Quickly I added, 'What about the cost of stamps, envelope and writing paper?' With considerable reluctance he chimed in, 'Well, give me three dollars and we'll call it even'. The three dollars was [sic] quickly in his hands. So, I got four lots, a hotel building, including a stable for about a dozen horses, for three dollars."[5]

As Johnson alluded to in his journal, the old building was in a serious state of disrepair. The crew at Gothic, however, was forever handy and they set about doing repairs themselves. It took about a month to fix the roof and the doors and replace window panes. The former hotel dining room was transformed into a laboratory; the former kitchen became the dining room; a previous cool storage for beer, wine, whiskey and "other liquors" was turned into the kitchen. Other rooms were carved out as an office, a storage pantry for foodstuffs, another storage room for their "meager" laboratory equipment, and some sleeping quarters for students. Despite the primitive and antique nature of the structure, it remained in use for eleven years, until 1940. At that time it was determined that it had become "too rickety for safety" and was torn down.[6]

The resourceful gentlemen and ladies of Gothic diligently looked ahead and decided that the better lumber from the building and the livery stables should be saved for future building projects, since funds were consistently limited. Indeed, these boards would soon be put to good use to build two student cabins near the Swallow's Nest. In addition, boards from the livery had been used in the construction of the "Bean Pod" in 1929.[7] Recycling is a venerable tradition at RMBL.

Garwood Hall Judd, for reasons of his own, seemed to have taken a fancy to John Johnson and his group of "bugologists". He was, of course, also looking out for himself. He may have procured the old hotel legitimately and sold it to Johnson for three dollars legitimately, but other transactions were a bit more "squirrely". He was all too happy to rent to the scientists and students properties he did not own. Some places he arranged to sell twice. Johnson himself commented years later, "[Judd] often sold mining claim stock to several different persons – using fancy paper to convey the same. He justified such sales as going to 'Eastern suckers' who should know better."[8] Judd obviously took pleasure in working with the scientists, and the scientists were pleased to take advantage of what he had to offer. Understanding that sometimes Judd shrewdly collected money he was not entitled to, Johnson said, "I harbor no ill will toward him. He was such a clever, tricky friend".[9]

Judd did arrange at least one other successful transaction for John Johnson. Three buildings had been constructed in Gothic in 1914 by a mining company from Ohio that was hoping to get mining going again. Judd managed to convince the company to sell these buildings to RMBL. One is no longer standing; the other two were eventually converted to relatively comfortable student dormitories. One of the buildings is known as the Ore House, named for its historical use as a place to process ore. The other is called Mammal Lab; for many years it was used as a laboratory to study mammals as a part of biological research.

In 1963, John Johnson asserted that, "Students learned just as much in those days as they do now with better accommodations".

Johnson was not one to waste time in pursuit of a goal. He knew he wanted the properties in the old town of Gothic, as well as the buildings that stood on them. The original deeds of the properties in question dated from 1882 through 1888 and had one thing in common – no one had been paying taxes on them. In the fall of 1928, Johnson petitioned the Board of County Commissioners of the County of Gunnison to allow the field station to acquire the abandoned lots. On August 10, 1928, the Board passed and adopted a resolution that the County of Gunnison, in consideration for the sum of $100, had "remised, released, sold, conveyed and QUIT-CLAIMED... all the right, title, interest, claim and demand..." to a total of 412 lots, all of which lay inside the town of Gothic. The sum total of these properties encompassed most of the town. A supplemental resolution was passed and adopted by the Board of County Commissioners on July 12, 1930, the purpose of which was to convey the properties to the field station, and a deed finalizing the action was filed and recorded at 4:45 pm on that same day at the Gunnison County courthouse. The failure of many to pay taxes resulted in the fortuitous acquisition of land for the growing field station.[10]

The acquisition of land and buildings, though vital to the long-term endurance of a facility, is only part of the picture of the creation of the Rocky Mountain Biological Laboratory. It is people who conceive ideas and actively work to turn those ideas into realities. So it was to be with Dr. John C. Johnson. and his merry band of idealistic biologists, who happened to love the great outdoors of the Colorado Rocky Mountains.

Truman Blancett: Gothic Pioneer

Truman Blancett was an early pioneer prospector in Colorado. He came to Colorado in 1862, at the age of 18, with his family. His father worked for the American Fur Company; his mother was a missionary. He himself grew to be a true frontiersman, traveling around the west and associating with others of his ilk, like Kit Carson and Buffalo Bill Cody. In 1878, he found his way to the area which would become Gothic, prospecting for silver. On June 25, 1928, Blancett returned to Gothic in time to be recorded as a visitor on the opening day of the new Rocky Mountain Biological Station.

Blancett would later write numerous articles about his life on the frontier, some of which were published by the *Denver Post.* Dr. John C. Johnson collected those articles and, in 1970, donated them to the Colorado State Historical Society, where they can be found today.[1]

In this letter of conveyance to the Historical Society, Johnson recounts his first meeting with Blancett.[2]

"On June 25, 1928 a squatter camped under some beautiful large Engelmann Spruces on the grounds of the Rocky Mountain Biological Laboratory [sic] at Gothic, eight miles north of Crested Butte in Gunnison County.

"Since this was the first day of the first year of this institution I was too busy as the Director, registering and housing our first students and research workers, to find out who the stranger might be.

"A visit to his camp the next morning proved to be most interesting and informing. The visitor was Truman Blancett, one of the earliest pioneers in the Denver region and possibly the first white man on record to visit the beautiful area soon to be known as Gothic City, later called Gothic. (It is very probable that prospectors and trappers did precede him but they left no written record.)

"Mr. Blancett arrived on horseback leading another horse that carried all his camping equipment and food for a couple of weeks. Of course, I never let him know that he was camping on privately owned property. Since he proved to be most cordial and 'talkative' I soon arranged for him to speak at our first Biology Seminar, which certainly was most appropriate.

"He said to me, and at this Biology Seminar, that he had returned 'to see the old place once again' where as a lone prospector he had camped out just a half

century ago. He said he arrived in September or October 1878, camped out under a very large evergreen tree and stayed about two weeks. He claimed that two weeks of digging yielded him about $600 worth of pure wire silver from a mountain valley that we now call Virginia Basin, overlooking the present townsite of Gothic. Some of the holes that he said he made can still be seen on the side of Virginia Mountain or Virginia Ridge.

"Mr. Blancett carried this silver back to Canon City on his pack horse, where he and other prospectors and pioneers often wintered. Since he had made such a rich find he claimed that he told only two other persons about his luck, who promised to tell no one else, but when he returned to the 'diggings' in Virginia Basin above Gothic there were already about 200 prospectors who preceded him in June, 1879.

"After this very interesting seminar at which he spoke he asked me if I would like to see some interesting 'documents' that he had written a long time ago, very likely before 1900. Of course I enthusiastically said 'yes'. And true to his word these 'documents', three in all, arrived in the mail addressed to me, postmarked July 30, 1928.

"I feel quite certain that the incidents depicted in the 'documents' are true and should become a part of the early history of the Denver area in the 1860's. Some of the experiences that he and his brother had can be and have been verified."

Johnson goes on to explain that he promised Blancett he would convey the "documents" to a library or historical society for safe and permanent keeping. However, Johnson himself moved east after that summer and lost track of the papers. In 1959, upon moving back to Colorado, he rediscovered them among some personal papers. In 1970, he conveyed the "documents" to the State Historical Society of Colorado, of which he was a member.

Johnson considered the arrival of Blancett an auspicious happenstance for the opening of the Rocky Mountain Biological Laboratory, bridging the prospecting foundation of the town of Gothic with the scientific foundation of RMBL in the same spot.

The First Five Years

True to his promise, Dr. John C. Johnson forged into 1928 and onwards as the sole financial backer of the Rocky Mountain Biological Laboratory. He had no administrative staff, but was a motivated, energetic, prolific writer. He actively sought both teachers and students for his new facility, and created publicity material and annual summer brochures. He was well connected in the field of biology and, although he didn't actually "do" scientific research himself, he knew enough people who did and who were happy to support the work at RMBL. He filled his years as biology professor during the academic year and administrator-in-chief in Gothic during summers.

Although a fairy tale start to the new organization was hoped for, there were almost predictable hurdles to overcome along the way. Evident in letters between RMBL Attorney Clifford Stone and members of the administration and trustees of Western State College is that the college was engaged in a push-back campaign against Johnson and RMBL.[1] There were several interrelated issues. During the summer of 1928, an effort was made by Stone, on behalf of Johnson, to acquire the rights to utilize the federally granted lands near Almont for RMBL researchers. The land had been granted by the Forest Service to Western for a "Western State College Biological Reserve" and had come to be known at the Rocky Mountain Biological Station. Johnson, as Dean and professor at Western, had been the driving force behind the activities at that reserve and, according to him, there did not seem to exist much enthusiasm for it on the part of other college faculty. Johnson did not want to lose access to the land, and he knew that the grant from the government stated that "this agreement will automatically terminate, if the said area is not used, being obviously abandoned, for three consecutive years."[2]

Attorney Stone, on behalf of Johnson and the new facility, requested that the land, or at least part of it, be assigned to them as part of the new venture. Vice-President R.W. Tallman's response was that other faculty members were very interested in the reserve and that, "Under the circumstances, it seems unreasonable and unethical to press the claim that the area has not been definitely assigned to projects and therefore should be sub-let to outside parties."[3] Despite statements in other letters from Stone to Johnson that he intended to continue to pursue the use of the reserve tract of land, the end result was that the land was essentially abandoned and was reclaimed by the United States Forest Service.

While continuing to claim research interest in the land on the part of science faculty at Western, E.M. Russell, Vice-President of the Board of Trustees, wrote to Attorney Stone and said "since it is the... college which... at the time of entering into the agreement with the Forest Service, would actually use the grounds... I would not deem it necessary for the college administration to grant any special permission to this well established part of the Institution to make use of the same." He then went on to point out that the organization incorporated within the college that was in charge of using the reserve lands was the "Rocky Mountain Biological Station". It is true that Johnson and his co-founders had incorporated their new facility in Gothic under that very same name and Russell focused on this. He accused the new organization of using its name to "undertake to connect itself with W.S.C. under various guises, even to the extent of trying to induce students to come", specifically by using W.S.C. "Office of the Dean" stationary to promote the new summer program in Gothic.[4]

Russell concluded his letter by saying, "It might be established that one of these Rocky Mountain Biological Stations is an infringer upon the rights of the other, and I feel that no official negociations [*sic*] could be carried on between them, nor between Western State College and both of them, until they are more clearly differentiated and their rights to exist more lucidly clarified."[5]

The name of the new corporation in Gothic was legally changed in 1930 to the Rocky Mountain Biological **Laboratory**.

Within this correspondence can be detected a strain of animosity, which could only have been directed toward Dr. Johnson. As previously pointed out, he was unceremoniously fired by Western State College. He always claimed this had nothing to do with insubordination or dereliction of duty, but was because of a personal dislike for him on the part of the president of the college. Attorney Stone pointed this out in a letter to Johnson conveying a copy of the letter from Russell. He wrote, "You will note that the letter reeks with personal feeling against you. It is my opinion that such letters as this are hurting Russell's position, and that he will hang himself." He went on to jot down in his own hand a suggestion that "our erstwhile friend is attempting to be clever", and perhaps if they were to apply to the "National Forest" to cancel the Western State contract, it would put an end to his cleverness.[6] There is no indication that this was ever done. On a happier note, the first year of operating in Gothic was a successful one. The *Gunnison News-Champion* of July 5, 1928 ran the following item:

> The old town of Gothic and environs has a population of close to 60 people this summer, according to Herbert Warfel, assistant at the biological station who was in Gunnison Saturday. Besides the students and faculty members and their families, numbering some 30, there are prospectors, engineers, geologists, cow hands and tourists, camped in or about the city. Mayor [*sic*] Judd seems to be in charge, as usual, and it seems good to have Gothic on the map again, if only for the summer season.[7]

Colleagues of the original founders of RMBL were among the first to support the fledgling field station with their presence during that initial summer. Among them were two from the University of Oklahoma who expressed their enthusiasm in letters to Johnson upon their return home in the fall of 1928.

Prof. Homer L. Dodge wrote that the week he and his wife spent at Gothic "has left a very bright spot in my memory and the fact that it has not been dimmed by the remarkably pleasant experiences I had throughout the rest of my trip [through Colorado] makes me certain that you have a remarkable location for the development of a biological station". While declining an offer of "permanent" residence in Gothic, he concludes with an offer of assistance in connection with the biological station saying, "It would be a great pleasure and privilege to have an opportunity to repay, in even so slight a way, for the splendid time that you gave [us] this summer". Dodge mentioned that, should he and his family decide to visit Gothic again, it would be nice to be afforded "some cottage" to live in, but he would be willing to consider setting up "a couple of tents, one of them provided with floor and low sides and a stove." What he really would have liked to do, however, was purchase "that whole farm down the river from the townsite".[8]

One of the co-founders, Prof. Aute Richards, wrote to Johnson in October 1928, detailing his travels home after his stay in Gothic that summer. While visiting friends and colleagues on this trip, he spread the word of what RMBL had to offer. He reported that after his return home to Norman, Oklahoma, the site of the University, he had not been in touch with many people outside his immediate community about the "Biological Station but in our own connection we have developed quite a bit of interest. A number of people have spoken about next summer and I feel rather sanguine about the situation."[9]

Richards went on to suggest some pro-active measures to get the word out about the Station, as he referred to it. It was his suggestion, along with unnamed others, that a circular be produced and sent to "all the people who we have on our list for special efforts in increasing the membership". He then suggested that it should not be "for general broadcasting" but available "for the professional biologists whom we think are interested". Typically, he thought Johnson should be the one to prepare the circular and oversee its distribution.[10] Johnson did just that.

The first circulars announcing each summer season touted the "Character of Area for Biological Work" as follows:

> The Gothic area comprises about a half million acres of virgin territory, uninhabited by residents, tourist or campers. There are numerous high mountains, rivers and streams, lakes, bogs, swamps, mountain and alpine meadows, and regions of perpetual snow. The great variation in elevation (7,000 to 14,000 feet) makes for great variation in plant and animal life. It is estimated that there are over 600 species of flowering plants alone, many of them unnamed. The animal species of the region have not been even estimated, or studied to any great extent. Doubtless here, too, are many new species.[11]

RMBL was chartered by the State of Colorado as a degree granting institution but has not ever offered a degree. It has, however, made credit available to students through their home colleges. Students could choose either to take two courses during the summer program, or one course and half time research. If they were truly serious about a project, they could opt to do research full time. The first and early brochures stated that, "each course is accepted by leading universities as equal to three semester hours," and that a maximum of six semester hours could be earned during the course of the summer. The brochures did not specify which "leading universities" accepted the work for credit.[12]

Courses were offered in the first few years by the founders of RMBL, as well as by "visiting" professors. John C. Johnson was in charge of courses on parasitology. L.A. Adams taught, or arranged to have taught, field zoology courses. Ecology classes were offered or overseen by A. O. Weese. Professor Paul B. Sears of the University of Oklahoma signed on to teach field botany. Aute Richards agreed to oversee "special courses", which were to be offered and taught by visiting faculty. In 1928, such special courses included lectures in Comparative Embryology, Genetics and Ornithology. Courses in all areas were not generally available each summer as they had to be assigned according to which faculty members would be in residence in Gothic.[13]

The cost of spending a summer at RMBL for students and researchers was minimal, based on the fact that the Lab had very little to offer in the way of amenities. Laboratory Fees for six weeks were $40.00; they included "no extras". Board ("excellent meals at $1.25 per day") was put at $50.00 for the summer. Housing – a "room or tent house (double) includes bedding and pillows, all new" – was a mere $20.00 for the season. Independent "investigators" were welcome and charged a fee of $25.00 for a research table; otherwise, their expenses were "the same as for students". The grand total for a student for a summer in Gothic, Colorado, studying at RMBL, was $110.00.

As housing was in great demand, the initial annual brochures encouraged biologists to erect their own cabins or cottages, stating that "sites nowhere excelled in beauty in the Rockies may be had for unbelievably low prices".[14] As it was not practical for many individuals to purchase land and build their own houses, a plan would eventually evolve whereby individuals could pay to erect a cabin in Gothic, on RMBL land, and be allowed to hold the lease on that cabin for twenty years, as long as they continued to do research on an almost annual basis. Eventually, those leases were permitted to be renewed for a second twenty years as scientists returned with their families summer after summer. After that time, the cabins were to be turned over to RMBL for use by students and researchers. Founder Aute Richards was the first member of RMBL to build a cabin for his own use. In October of 1929, after the second summer session, he wrote to Johnson that, "We should very much like to get our own living quarters under way so they can be used by the time the session begins".

Richards was quite serious about his housing plans. In a letter to Johnson dated May 4, 1930, he outlined his thoughts about the project and included a hand-drawn sketch of how the cabin should look. He also broke out estimated costs. He said he would like it to be below the spring, so that it was near running water, but "it would need to be in a dry spot where we could get the car up to it". While waiting for acquisition of the larger building(s) belonging to the Disette brothers, "we would have just a cabin with the second floor for sleeping". He planned to do interior finishing himself so as to minimize the cost of labor.[15] His financial calculations, including logs, sheeting, flooring, windows, doors, nails, stairs, etc., plus labor, came to $333.75. Dr. Johnson amended that in his own hand to $341.00. It appears, from later notes, that the cabin, which measures twenty-four feet square and eight feet high (to the eaves), cost closer to $450.00.[16]

Dr. Johnson's handwritten list of first day items to cover with students and others working in Gothic in 1929 would become the norm for his annual first day welcome.

1. Welcome – virgin field…
2. Here for study & research first – recreation second
3. Rules – None except to be a lady or gentleman
   If in doubt, don't do until ask Mrs. Johnson or myself
   Tell us
4. Courses at 9 am & 2 pm
   AM – F. Botany; Parasitology
   PM – Ecology; Protozoology
5. Field Botany for most people
6. Fees – this afternoon or tomorrow
   No extras except… personal books
7. If books – materials – order soon
8. Talk around
9. Tell me what courses
10. Not responsible for accidents
11. Help you

In 1930, he added: "Complaints – Tell us. Mrs. Johnson & I. Don't complain much."[17]

Money – and the lack of it – was an issue of continual concern for Johnson. He kept careful and copious records of every penny spent. Responsibility for payment of all bills was in his court, per the agreement he had made with his co-founders. In 1929, he recorded the income from Laboratory Fees as $611.67, minus $15.00 for "4 days work for bracing, closing, cleaning laboratories, etc." This amount reflected $40 fees paid by fourteen attendees, plus one discounted fee of $26.67 and one research table fee of $25.00. On the 19th of July that year, he recorded estimated expenses as more than twice that amount. Itemized expenses included instruction, books, food, milk, insurance, coal, gas, oil, telegrams, bulletins, supplies, permits and more.[18] Some of these, such as food, fuel and milk, would be recouped through fees for room and board. But there is no doubt that the station operated on a shoestring budget.

Although the Laboratory had been able to acquire much of the land encompassed by the town of Gothic, it was situated in the middle of national forest land. Therefore, the problem of no control over the surrounding environment presented itself fairly early. For the most part, early correspondence indicates that the national Forest Service was happy to have RMBL for a neighbor. In a letter to Johnson dated May 13, 1929, Gunnison National Forest Supervisor Theodore Krueger wrote, "I am very glad to note that you will again be at the Rocky Mountain Biological

Station this summer, and I assure you that we will be only too glad to cooperate with you to the full extent in making the resources of the Gunnison National Forest available to your Station."[19] He also expressed his hope that students at the Station would become interested in helping to solve "a number of interesting biological problems" that had developed.[20]

Protecting the pristine nature of the environment around Gothic was extremely important so that the scientists' work would be unpolluted and unencumbered by careless or unknowing passers-through. Richards, Weese and Johnson discussed in 1929 the prospect of requesting that Gothic be withdrawn from fishing, which would require permission from the Forest Service. In January 1930, application was made to the Chief of the Forestry Service in Washington, D.C. by W.B. Bizzell, President of the University of Oklahoma, to "withdraw the area immediately surrounding this station from camping and fishing… if it is to function as a Biological Station, it will have to be completely protected from transient visitors to the area". Conversations about this topic resurface repeatedly in correspondence among the trustees throughout 1930.[21]

Quite opposite to protecting the area was a problem that arose pertaining to acquiring a sufficient number of "specimens" for research. In response to an inquiry by Johnson to the State of Colorado Game and Fish Department about collecting birds for scientific research, he was told that the law required he "obtain the endorsement of an incorporated society of natural history of Colorado, also required to be under bond of $100.00, and fee for permit is 50 cents". And, he and his people were reminded they were prohibited at all times from "taking birds protected by the federal law as well as the state law".[22]

Johnson, Richards and Weese kept up a lively stream of communications among the three of them in the early years. There was constant discussion about who to recruit to teach courses, where to recruit students, where would money come from to pay for things (insurance was very expensive) and how would facilities be built and otherwise acquired. In his recommendations to the Board of Trustees in 1930, Johnson offered specific suggestions concerning the acquisition of property. A property with a cabin on it was purchased from a Mr. Price in Denver for $250. RMBL had already been granted "tentative" use of property owned by the Disette brothers of Cleveland, Ohio, but Johnson was negotiating for the acquisition of it. When Garwood Hall Judd died in May of 1930, Johnson suggested that the RMBL trustees "investigate the appraisal value of the property" and that steps be taken to purchase all or part of it, if reasonably appraised.[xxiii] He notes in 1930 that since the previous summer, one log cabin had been built, one frame house had been bought for laboratory purposes (Price Cabin) and two small student cabins had been built, which would accommodate three or four additional students. "Members of the Station now own eight buildings – a laboratory and seven buildings used as residences or lodging places. It also uses one large building belonging to the Disette estate." The Disette brothers owned a mining interest in Gothic and there were three buildings on their property. One of these fell into disrepair; the other two were used as laboratories for a number of years, then remodeled and converted into student dormitories.

Johnson was also pursuing the use of additional areas of land for pristine biological research. He suggested in 1930 that the secretary of the Board of Trustees write to the President of Western State College "demanding [*sic*] official permission to use the area of some 33 acres located about four miles north of Gothic, known locally as the Biological Forest and Research Area". He also thought it would be prudent to contact the U.S. District Forester in Denver and ask that Experimental and Research Reserves in the vicinity of Gothic be put aside. He was very specific about the areas in question. The area "one mile south of Gothic and extending to Schofield and Maroon passes (including the East River and Copper Creek watershed and their tributaries" should be set aside for "Biological Investigation and Research". Also, he wanted the land between the East River and Gothic Mountain either donated to the Station or placed under its control, "to insure a perpetual unpolluted water supply for the laboratory and houses in the village".[24]

It is obvious that Dr. John C. Johnson had great plans for his fledgling organization. And, he and his trustees were quite successful at convincing those in power that they were right.

In February of 1931 a letter was sent to Aute Richards by the Regional Forester in Denver, Allen S. Peck, informing Richards that the Regional Investigative Committee had met in January and had discussed how the Forest Service and the Rocky Mountain Biological Station could work together. In addition to research on issues of mutual interest to the two organizations, the Forest Service had agreed to set aside an area containing 509 acres of timber type and 395 acres of grassland, brush and barren land. The purpose of these reserved lands was, "Permanently to preserve in an unmodified condition areas representative of the virgin growth of each forest or range type within each forest region so far as they are represented within the National Forests, to the end that its characteristic plant and animal life and soil conditions, the factors influencing its biological complex, shall continue to be available for purposes of science, research and education." He also went on to say that they would be able to accommodate the interests of the Station in that "no summer homes or commercial uses will be permitted on the east slope of Gothic mountain, which constitutes your watershed, or in the upper Copper Creek basin". Peck reminded Richards that the Forest Service had no control over mining or fishing, nor would they desire to restrict public passage through the area. However, he felt that traffic through and around Gothic and the reserved lands would not be an issue if the lakes and Copper Creek were not stocked, "as fishing is the main attraction".[25] The trustees and members of RMBL were elated by this support.

As a way of building continuing support for the Rocky Mountain Biological Laboratory, the Trustees decided to offer memberships in the corporation, for an annual fee. Dues for 1930 were $2.00. Members were not required to be in residence each summer, nor were they even required to be scientists. They just needed to exhibit the interest in the Lab that would make them agreeable to paying annual dues. In 1930, twenty-four individuals were listed as Members. These included the original five founders, some of their spouses and academics from a variety of colleges and universities in Colorado, Oklahoma, Illinois, Kansas and Pennsylvania. Students came to study from all over the country, including Oregon, Maine, Iowa, Illinois, Oklahoma and Pennsylvania.

Enthusiastic students unfamiliar with life in the Rocky Mountains wrote eagerly to Johnson, as Director, asking simple questions about what to wear, how many suitcases to bring, what about books… He answered each and every one of them, individualizing his responses rather than creating a form letter. As to clothing, "Any type of clothing is in style – from hiking outfit to typical college clothing, perhaps the former dominating. I usually wear a hiking out-fit with high laced boots". All bedding was furnished, but one might want to bring along a few extra towels. Definitely bring your own microscope; it could only be better than the ones provided. Text books would be available to use. Two suitcases were preferable to a trunk. And, "you will receive a hearty Western welcome. Autos will be at the railroad station… to take you to the Laboratory without extra cost".

Not everyone could afford even the meager tuition and fees for the summer, hence the tradition of students and faculty members as laborers in Gothic began quite early. One student wrote to Johnson in December 1930, wishing him and his family a happy Christmas and suggesting that, as much as he would like to return in summer 1931, "paying back borrowed money is a slow process". He went on to ask if any new cabins were to be built. It seems his father was a carpenter and he had worked alongside him in the trade. He offered to work in exchange for part of his expenses, saying "it would help out wonderfully". Wielding hammers and saws, cutting and carting logs and doing whatever was necessary to build, rebuild and maintain the physical plant of Gothic helped many a student afford a summer at the Lab.

Reacting to her time there, an advanced student wrote to Johnson after her summer in Gothic that, it had "built up a rested nervous system for the next ten strenuous months". She had been able to take only one course, in ecology, and requested those credits be sent to her home school. Speaking of the experience, she wonders that Dr. Richards disapproved of a session spent on only one course, his reasoning being that it would be best to have no spare time in which to get into mischief. But, she wrote, "it is perfectly plain that if one is inclined to get into mischief he (or she) will find the opportunity to do so, by burning the candle at both ends. Don't you think so?"[26]

As the summers passed, the reputation of RMBL grew and spread. Faculty and students alike flocked to the little community high in the Rocky Mountains. All came to do research or take courses, but there was also an undercurrent of devotion to the place and to the people that began to develop. Some came only once, but many would return annually for years or decades, or come back after a hiatus elsewhere. Actual numbers in attendance during those early years vary depending on the source. Johnson's hand written records put 1931 at twenty-one students and five staff. He later amended that to "twenty-eight investigators, instructors and advanced biological students coming from fourteen different states… and two foreign countries", Sweden and Greece. Seventeen colleges and universities were represented in the list Johnson created for his records. It is probable that some individuals had family members with them who would have accounted for an increase in numbers. The *Gunnison News-Champion* of July 16, 1931, however, reported that there were "43 enrolled students and a number of faculty members in residence at the Rocky Mountain Biological Laboratory at Gothic". The newspaper article went on to report that, "they are a high class group of college men and women, only several of the 43 having anything less than a master's degree". What is certainly true is that 1931 was a successful year for RMBL.

Johnson's Director's Report written in November 1932 marked that summer season as having been "another successful one – in spite of the prevailing financial difficulties". He put the total number present in Gothic at forty-four, which included investigators, students, staff and family members. Most of the students had already earned a Bachelor's or Master's degree, and ten held a Ph.D. In his own hand, Johnson recorded that, "The corporation (a non-profit organization) now owns some seventy acres of land in the Gunnison National Forest on which are located some fourteen buildings, including both cabins and laboratories. Two buildings were added last year, a Mess Hall (purchased) and a new Director's residence." It was hoped that a bath house and a new research laboratory would be completed the following summer, but sufficient funds had not yet been raised.

Significantly, in the initial five years of its existence, the Rocky Mountain Biological Laboratory had thrived and laid down roots for the facility it would become. Unfortunately, John C. Johnson would face that sixth summer in a manner he could never have anticipated. The initial joyous discovery that his wife and partner in everything was once again expecting a child turned sad and tragic when it was learned that there was no baby, but a tumor growing out of control. In a day and age when the only way to know what was happening was to wait until the baby was born, the cancer had spread to become fatal. Letters of support and love arrived for both John and Vera Johnson, repeating the hope that all would be well and Vera would recover quickly. Perhaps the most touching note remaining to us came from Volita Wells in Philadelphia, Pennsylvania. In November 1932 she wrote:

> I was very sorry to hear today that Mrs. Johnson is ill. I always felt in our Educational Biology classes that even if the students forgot a few dozen scientific names, they would never forget the picture of what married life in its fullest meaning can be. Some will always enjoy Biology, some never will, but there are few who are not interested in marriage. There seems to so little realization of the growth of happiness in marriage – some of the young folks seem to think that it ends with the honeymoon – that I think it of inestimable value to be given glimpses of the beauty of family life that has grown.[27]

Vera Adams Johnson passed away on April 6, 1933 and was buried in the family plot in Gunnison. The Rocky Mountain Biological Laboratory continued without her, but her role as a founder, "Advisor to Women" and hard-working supporter has never been forgotten.

Who Is RMBL?

An organization exists as a concept with brick and mortar buildings to contain its people and materials. It is the people, however, who fuel that concept, breathe life into it, cause it to function. In the case of the Rocky Mountain Biological Laboratory, the person who conceived of it and those persons who funded and physically "worked" it and contributed ideas that made it function were the lifeblood of the organization. Mother Nature certainly played Her part in providing the venue, but Nature exists in a world of its own making, a world that incorporates the land and sea and sky – as well as the people who inhabit that land, sail the seas and breathe the air from the sky. Gothic had existed as a place in Nature long before any Native American or prospector or entrepreneur set eyes or foot upon it. It was the people who came flocking to that little place called Gothic, Colorado beginning in 1928 who are RMBL.

The founding members of the Laboratory were the ones who sculpted a curriculum for summer study and research, taught the courses or found colleagues to do so, wielded saws and hammers to build the cabins, handed out blankets to the attendees, saw to the purchase of food, fuel and laundry services, worked in the kitchen preparing meals, communicated with the outside world and connected with the local community in Crested Butte to find help to get things done. But this could last only so long. Time passes and people move in different directions. Dr. John C. Johnson was intensely focused and dedicated, but he became a widower and had to care for four young children. Even after he married, he needed to incorporate his new wife into life in Gothic. His focus never wavered, but reality imposed challenges. Despite his good intentions and even after constructing a cabin, Dr. Richards did not dedicate himself to annual summer stays in Gothic. Dr. Weese was almost as dedicated as Johnson but he, too, spent increased time organizing from a distance. It was clear that the actual hands-on teaching was going to have to come from others. As the purpose of the organization was to encourage biological research and education, researchers and educators needed to come.

Come they did - men and women alike. An interesting observation about the Rocky Mountain Biological Laboratory is that it was, from day one, a cooperative effort not just between colleagues and friends, but also between spouses. Perhaps it was Johnson's reliance on the help and support of his first wife, Vera, and then his second wife, Mildred, that set the stage for RMBL as a place where married couples were welcome as equals in research. In a time when women were overlooked in the professional world, they came to Gothic and were treated with respect. They were encouraged to pursue their own research and not just work as laboratory assistants to their husbands. The importance of women as contributing individuals began in the early years of RMBL and has continued throughout its existence.

With no disrespect to the men who accompanied them, a word must be said about the amazing women scientists at the Rocky Mountain Biological Laboratory.

Women, Science and Gothic - The Early Years

Science was not a professional field that encouraged the participation of women in the early twentieth century. It is not that women didn't study the sciences and pursue degrees in the various fields of study. Those who attended college were expected to take courses in the liberal arts and sciences in order to earn their baccalaureate degrees, as were men. Graduate degrees became a little more challenging for women and specialization in a field did not guarantee employment in that field. Most likely, employment in a science-related job was of a secondary nature. The coveted professorships went to men, more often than not. Women made excellent laboratory assistants, or part-time teachers, no matter how educated they were. This is where a place like the Rocky Mountain Biological Laboratory made a difference. It not only allowed, but encouraged, women to do field research and to teach courses in the summer program. It is a place where a young woman of science could blossom and satisfy her desire to learn and to teach, as well as to make an impression on young students in the field.

Mildred Hoge Richards, wife of RMBL Founder, President and Trustee Aute Richards was one of the first "women scientists" to teach and do research at the Laboratory. Mildred was a geneticist and zoologist, educated at Goucher College and Columbia University, where she earned a Ph.D. in genetics in 1914. Prior to her marriage in 1917, she worked as an instructor in zoology at Indiana University. Despite her education and professional teaching experience, she was not offered a position at the University of Oklahoma for almost all of the thirty years her husband taught there. At long last, in 1947, she was given an appointment as associate professor. She held that position for only one year before retiring in 1948. Whether her professional career was stunted by nepotism or bias against women in science is unknown, but Mildred was able to fulfill her interests in science by working summers at the Rocky Mountain Biological Laboratory. Not only was she a member of the corporation in her own right, she did research and taught courses in genetics. She also raised two sons. She did not publish in her field, but did author and publish a book entitled, ***The Descendants of Samuel Hogg of Wilmington, Delaware*** (Tucson, AZ: 1959).[1]

One of the first professional couples to accept a role at RMBL was B.D. and **Harriet Barclay**. Dr. Paul B. Sears, head of the botany department at the University of Oklahoma, had been slotted to teach at RMBL in the summer of 1929. When illness forced him to withdraw, he recommended that Dr. Bertram Donald (B.D.) Barclay from the University of Tulsa be tapped for the position. Copies of telegrams between Dr. Johnson and Dr. Barclay dated June 1929 confirm that, although the Barclays had "just driven to Minnesota for the summer", B.D. was willing to accept the position if the conditions were "satisfactory". Johnson responded by offering to pay living expenses for Barclay and his family at the station, plus forty dollars for traveling expenses. He added, "We have chef. Wife need not cook. Required to teach only one class – field Biology. Takes one-fourth instructor time. Rest of time free for research, study or recreation". There was no offer of a salary, nor was one requested.[2] The *Tulsa Collegian* reported in July 1929 that Dr. Barclay would be working at RMBL and would be assisted by his wife, Dr. Harriet G. Barclay, "who is also a field botanist of note".[3]

That summer of 1929 was the beginning of a relationship with RMBL that would last throughout the Barclays lives. Even after B.D. was tragically killed in a car accident in 1953, Harriet continued her connection with Gothic, bringing students to study and do research, passing the torch to younger generations.

Harriet George Barclay (1901-1990) was born and raised in Minnesota where she received her pre-doctoral education. She moved to Chicago to study at the University of Chicago, where she earned her Ph.D. in 1928. There she met and married Bertram D. Barclay, who was asked to organize the nascent botany department at the University of Tulsa in 1929. They moved to Tulsa and while Bertram created, ran and taught in the department, Harriet also taught, raised their two sons, spent summers in Gothic collecting and preserving high mountain flora, and studied for her degree in art so she could showcase it in full color. During her very first summer in Gothic, she began an herbarium of plants of the region, which has been continually expanded and maintained over the years.[4]

Barclay claimed that her life-long passion for natural science began as a child, when her family would go on outings into the woods and natural areas of her native Minnesota. She was absorbed by the lushness, color and unique features of the wildflowers in the Colorado Rocky Mountains. While working her way from lecturer to full professor of botany at the University of Tulsa, she also pursued a bachelor's degree in art. She was known to be an excellent painter. It was her goal to showcase the natural beauty of the plants she loved in all their glory. Since plants shrivel and turn brown once harvested, she opted to paint them so that her students could study the real thing in the herbarium side by side with the replication of what it had once looked like. She was known to be an excellent painter and some of her art remains at RMBL and in the hands of individuals associated with the organization. When color photography burst on the scene, Harriet embraced the camera and taught herself to become an accomplished photographer. In the words of someone who knew and worked with her, "first with painting, then with photographs, Harriet enthused generations of college undergraduates, then members of various study groups, then townspeople" with her love of plants.[5]

When B.D. died, Harriet was named head of the botany department at the University of Tulsa. She remained in that position until 1972, when she reached mandatory retirement. During the years with and without her husband, she actively encouraged students to pursue botany or work in other areas of ecology. According to one of her students who was perhaps her most ardent supporter, Dr. Paul Buck, the classes she taught at TU produced "Ph.D. botanists all over the country".[6] Buck himself was inspired by Harriet to work in the field, and visited Gothic the first time in 1958 as her student. He went on to teach at the University of Tulsa, travel the world with Harriet and was curator of the Harriet G. Barclay Herbarium.

It was after the death of her husband that Harriet began her international travels. The beneficiary of a 1958 National Science Foundation grant, she spent a year and half in the late 1950s traveling through the high Andes in Venezuela, Columbia, Ecuador and Peru, doing botanical field work. She credited her work at Gothic above timberline with inspiring her to go to South America.[7] She would spend the rest of her life traveling to every continent except Antarctica – and always regretting that she never made it there. In a 1977 Christmas card she spoke of "too much" happening to permit her sending cards the previous year, and recorded having gone around the Horn of South America with the Smithsonian in February and March of 1976, teaching at RMBL during that summer, traveling to Costa Rica in December and attending her induction into the Oklahoma Hall of Fame. "That event, in November, at a dinner for 1,700 people, was unbelievable and thrilling!" She describes the summer of 1977 has being divided between teaching in Colorado and guest lecturing for the Smithsonian on a trip to the British Isles, "both large and small". She also visited Maine and planned to go to South America in 1978, on her eleventh trip, "perhaps not my last". Two years before she died, *Modern Maturity* magazine reported that, "She's hiked Kilimanjaro, trekked New

Zealand's Southern Alps, and paddled (quickly) past crocodiles in the Amazon". She had just returned from New Guinea and was planning that trip to Antarctica. "I'm in a hurry. After all, I'm 86", she is quoted as saying.[8] She was a lady on the go!

Harriet Barclay never published a research paper. As a scientist, she tended toward the roles of explorer, discoverer, cataloguer and teacher. According to Bobbi Peckarsky, a RMBL scientist who first came to Gothic in 1974, "her treasure was in her teaching". But during the course of her various explorations, she discovered at least nineteen species designated "new to science", many of which incorporate "Barclay" into their scientific names. One South American variety was dubbed "Senecio Harrietae".[9] In 1969, she was the motivating source behind the creation of an 85 acre nature preserve in Oklahoma, northeast of Tulsa, called the Redbud Valley Nature Preserve, which later built the Harriet G. Barclay Nature Center on site.

The Rocky Mountain Biological Laboratory is proud of its thirty-five year association with Harriet Barclay and the feeling was apparently mutual, as she named it as one of the places donations were to be made in her name when she died. Many scientists recall fondly their brief meetings with her, or share stories of their families' long associations with her. The RMBL legend of the woman is steeped in tales, two of which are repeated frequently and have at least partly to do with her height. Harriet was a tiny woman, only 4'11" tall. B.D. was not a tall person, either. When it became apparent they would be spending summers in Gothic, they decided to build a cabin. They hired Joe Savoy, a Crested Butte builder/handyman who built or helped to build many a cabin in Gothic. He scoured the surrounding mountainsides for logs that had survived weather and a forest fire for his projects. Custom built to their specifications, the two-story Barclay cabin accommodated Harriet's small stature, even down to the placement of windows. According to a request for funds to help rebuild the cabin in the autumn 2008 RMBL Newsletter, "Anyone over 5'5" tall must duck to get a view from the cabin's windows. The west side of the cabin still has the original chinking between the logs and bares dated signatures of many of the Lab's original members and their children". Some people, now adults, who remember summers in Gothic as children also recall swinging in the hanging "bed" in the middle of the living area of the cabin. The cabin was restored in 2009 at a cost of around $175,000. As it was an historic property, architectural review was part of the process, adding to the details that needed to be addressed. The man who did the work, local contractor Kevin Donovan, said that some consideration was given to either raising the ceilings or lowering the floors to provide for greater headroom, but in the end the cabin was left essentially as it had been built. It continues to be used annually by researchers and students, who must be careful not to bump their heads.

The other oft remembered Harriet Barclay story has to do with her driving. Each summer, Harriet would declare it was time for a field trip. She would pile students into her oversized car and head off into the mountains in search of flowers and plants. She could barely see over the steering wheel, but would drive at breakneck speed over not-very-good roads, screeching to a halt when she spotted an interesting specimen. On one such trip she decided to stop in Paonia, across Kebler Pass, for a break. Her dusty students piled out of the car and began wandering the streets, in and out of shops. They were such a motley-looking group that the town police asked them to leave, believing they were drug dealers.[10]

Perhaps Harriet Barclay's greatest legacy to RMBL is the herbarium she created and maintained during her years there. The specimens, dating back to 1927, are kept in filing cabinets in a designated collection room. Students and researchers have access to them and are encouraged to handle them carefully and replace them properly. Unfortunately, mistakes happen, as do accidents, and the collection may not be as perfectly intact as many would like. Professor Rick Williams from Idaho State University is curator of the herbarium and is working to see that it is preserved, both in fact and online. Information on the collection of 5,264 specimens is available at *intermountainbiota.org*

Many of Harriet Barclay's students found their way to RMBL. Some returned; some did not. But a student of note who made her mark was **Jean Langenheim**. It is interesting that, though taller than Harriet and B.D., Jean and her husband, Ralph, rather than build a cabin, famously made their home in Gothic in a chicken coop. But more on that later.

Jean Langenheim was born in Homer, Louisiana in 1925 and grew up enjoying train travel with her mother as they visited family in New York and other eastern states. Her father, who worked for the railroad, lost his job as a result of the Great Depression and the family relocated to Tulsa, Oklahoma, where he found work as an accountant with a truck line. Jean's formative years were spent in Tulsa and she chose to attend college at the University of Tulsa. It was there that she met Prof. Harriet Barclay. Jean's interest was in geology, but she was discouraged from majoring in that field because, "women were not allowed to take a summer's required field course."[11] At that time in history, it was not thought to be appropriate for women to live in the field with men, thus creating a serious problem for those women who were interested in field sciences. During her freshman year at TU, Harriet Barclay suggested that Jean's dilemma could be resolved by combining geology with botany, thereby completing her field work requirements with Harriet. She enrolled in courses in both fields; a career and a lasting friendship were hatched.

Both Barclays were involved in courses offering field trips and Jean was able to immerse herself in field work through these classes in the summer of 1944 and spring of 1945. The inveterate Barclay team, despite war rationing, "managed to obtain gasoline and cars with sufficiently decent tires for field trips around Oklahoma".[12] In addition to being able to complete her field study requirements, Jean began to appreciate how personal and professional relationships could be built and nurtured through the field study process.

Just prior to her graduation from TU in 1946, Jean met her future husband Ralph Langenheim, a student in the master's program in geology at the University of Colorado. As an engagement present, he gave her a geology hammer, which would be "symbolic of a partnership to work together".[13]

It was during the summer of 1946, with the encouragement of Harriet Barclay, that Jean Langenheim attended her first session at the Rocky Mountain Biological Laboratory. It seemed fated that she be there, as so many people associated with the University of Tulsa had ties with the place, as did her future husband, Ralph, whose family had coincidentally summered in Gothic. She took Harriet Barclay's Field Botany course, as well as a course in Field Ecology, taught by Barclay and A.O. Weese. Her personal recollection of this summer is a fond one.

> The Gothic area is not only an incredibly beautiful environment, but also an exhilarating place to study and do research. There was no end to fascinating discoveries in these mountains, expedited by knowledgeable people who also loved the area. I enjoyed the comradeship of the students, who were a mixture of friends from TU and others from Harvard, Syracuse, Georgia, and Oregon State. It was the first time for all of us to live in such primitive conditions. We shared in good humor the jobs of hauling water for bathing and wood and coal for heat, taking care of kerosene lamps for light, and so on.[14]

Graduation and marriage saw Jean and Ralph Langenheim living and pursuing graduate degrees in two different states; Jean in Minnesota and Ralph in Colorado. Summer in Gothic was where they were reunited before both settling on the University of Minnesota to pursue their Ph.D.s. Jean used the Gothic earthflow as the basis for her master's degree, and she and Ralph fenced a weather station in Gothic in 1947. She had been using the area to assess some microclimate conditions and it needed to be protected from wandering, free-range cattle. As they both realized that RMBL would be their base for summer field research while pursuing their doctorates, they decided that summer to lease land from the Lab on which to build a cabin. They chose a plot above the base of the town, amid some lovely aspen trees. The harsh financial realities of life as graduate students caused the young couple to accept the fact that their dream of a cabin in Gothic might be premature. Needing an affordable place to live there, however, they hit upon the idea of purchasing a prefabricated domed chicken coop from a catalog. The coop was delivered to Crested Butte via the narrow gauge railroad that operated there at the time, and was hauled by truck to Gothic, where they assembled it. The coop came with a wooden floor and cost a grand total of $200. The young Langenheims also purchased a bunk bed that fit against the back wall, and built a table and stools to fit beneath one of the front windows. They further furnished their makeshift home with a sheepherder's stovepipe stove, shelves and homemade curtains. A nearby outhouse completed their necessities.

The Langenheim chicken coop is but a memory anymore, but it was utilized for decades after Jean and Ralph were done with it by students who were happy to tell tales of their time there.[15]

Research in the Elk Mountains in which Gothic and the Rocky Mountain Biological Laboratory are situated could be challenging. The summers of 1948, 1949 and 1950 were spent mapping hundreds of square miles of the region. Jean and Ralph would camp out for days at a time, sleeping in a small tent they carried and trying to stave off the nighttime cold by layering sleeping bags. Water came from melted snow or streams, meals consisted of dehydrated foods or cereals cooked over open fires. Treats carried with them included dried fruits, cheese and chocolates, but nature provided fresh wild strawberries, huckleberries and gooseberries. Jean kept pace with her husband, never dreaming of giving up life in the wild and the adventures of field work. Or, almost never…

Recounting the survey work they did, Jean recalls that, "Frequently I was the one holding the surveying rod and I was nearly blown off more than a few alpine ridges. I also had a few literally hair-raising experiences, holding what could have easily become a lightning rod during approaching thunderstorms".[16]

Animal watching was part of the fun, and the Langenheims reveled in sighting squirrels, chipmunks, marmots, pika, mule deer and elk. An occasional porcupine crossed their paths and, in the alpine areas, ptarmigan with their young sometimes appeared.

Returning to the coop was a welcome relief after days in the harsh field environment. Basic as it was, it was home for the summer, and home comes with its comforts. A warm, dry environment, hot meals often prepared with a pressure cooker, occasional fresh trout from the local creek and hot biscuits baked in their own stove were relished after the routine of cold nights and dehydrated foods. There is much to be said for being warm, dry and well fed.

Summers in Gothic came to a halt for a while, as first Ralph and then Jean completed their dissertations and moved into their careers. Or, perhaps, one should say, his career. Ralph accepted a position at the University of California, Berkley in 1953 and Jean went along as his wife, but she was given the position of research associate the next year. They traveled together on a National Geographic expedition to Columbia in 1953 and she was hired as a "paleobotanista", to study plant fossils.[xvii] Jean points out in her memoir that her 1953 passport was "evidence that women international travelers were considered as adjuncts to their husbands", as his passport said he was "accompanied by" (fill in the blank): wife or minor child. Her occupation was listed as "housewife".

Living in the San Francisco Bay area offered Jean the opportunity to work away from the confines of Berkley. She relates that, "… teaching, especially at women's colleges, was a major professional outlet for women with advanced degrees in the 1950s". There were two women's colleges in the Bay area, but her first big opportunity came in 1955 when she was asked to return to RMBL as a faculty member and teach the Field Ecology course. She gladly left Ralph to his own summer commitments and headed to Gothic in their new Chevy sedan (which she planned to use for field trips).[17]

Jean arrived in Gothic in the summer of 1955 to a changing community. Mining in the region, its original cause for being, had grown essentially dormant, and the narrow-gauge railroad was being dismantled. The small town of Gothic was being electrified, thanks to a grant from the National Science Foundation. The hitch was that the electricity was connected only to the research facilities and administrative offices. Housing was still primitive, but it was considered to be a part of the students' "mountain experience". Students were also expected to haul coal and wood to their cabins and the classrooms and generally participate in the running of the place. But they were treated to hikes to places most would consider inaccessible, often assisted to trailheads by personal vehicles of the faculty. Those vehicles were very much the worse for wear by the ends of the summers, as the roads were mostly ruts, potholes, rocks and mud.

Jean returned to Gothic to teach Field Ecology a second time in 1959. Ralph had lost his position at UC Berkley and they were headed to the University of Illinois at Urbana. Once again, Ralph had other plans for the summer, but he joined Jean, along with her mother and some family friends, for some time at the end of the season. They made a long anticipated trek to Aspen, over Conundrum Pass, in order to see an opera. Jean records that she did not find this summer's work at RMBL as satisfying as it had been in 1955, perhaps due to differences in the students, perhaps to her own sad state of mind at having to leave Berkley. She would return for one more summer, in 1962.[19]

The Langenheims divorced in March of 1962. Jean drove to Gothic that summer with an understanding that she was beginning a new era in her life, and returning to a place that held so many memories. She and Ralph had begun their married life there; she had worked productively on her dissertation and, later, teaching undergraduates. This time, teaching the Field Ecology course would bring her much satisfaction. She relates that her students were enthusiastic and well-prepared, and helped her to "project my thoughts forward rather than backward". This was to be her final summer teaching at RMBL, but she remained involved in the Laboratory for several years. She served on the Board of Directors from 1962 to 1966, as vice-president from 1965 to 1966. Her visits from that time on were just that – visits.[20]

Jean Langenheim went on to build a very successful career in her own right. She traveled the world in the name of research and retired from the University of California in Santa Cruz in 1994, after having returned to California to work there in 1966. A woman of science who forged through the growth of feminism and stood much of her life on her own two feet, paving the way for others to follow in her path, she is well-remembered at RMBL. Her commentary about the area and her reasons for wanting to work there is poetic.

> I loved the striking colors of plants characterizing the vegetative zones, which only added to the color backdrop of the area's rocks. Their interplay seemed to me like the brilliant contrasts and light effects rendered by late nineteenth century Impressionists such as Monet, Pissarro, and Gaugin. And if landscape color was not enough the abundance and array of colorful individual plants also enchanted me. Simply put, it is an exceptionally beautiful place to do research.[21]

It is not possible to write about women of science at the Rocky Mountain Biological Laboratory without giving mention to Dr. **Theo Colborn**.[22]

Theo Colborn was born and raised in the eastern United States and, along with her husband, studied at Rutgers College to be a pharmacist in the 1940s. Together they opened a series of pharmacies in New Jersey, including the first in-hospital pharmacy in their area, before becoming disenchanted with the lifestyle. The hours were grueling and took a severe toll on family life. Also, Theo began paying attention to emerging environmental issues in the 1960s and was concerned about the quality of the air and the water in the northeastern United States. She and her husband began to talk of moving their family west, to start anew. In 1962 her husband answered an ad in the New York Times from a German family who had brought a Volkswagen "bus" to the United States to use for a cross-country camping trip. They were returning to Germany and looking to sell the vehicle. Her husband drove to New York and bought the bus on the spot. That summer, they packed up three of their children and drove cross country on "an adventure". Arriving in Boulder, Colorado, they decided they had discovered their new home. After putting their affairs in order in New Jersey, they settled into life in Boulder as students and teachers. This new life did not suit her husband, but Theo quickly realized that she wanted to study environmental issues, particularly freshwater ecology.

Life in Boulder rapidly gave way to a move further west to the area around Paonia, Colorado. Having purchased a farm, Theo settled into actively running it even as her husband became increasingly ill from what would today be recognized as post traumatic stress disorder resulting from his military experiences in World War II. She worked night and day to make a go of the farm, moonlighting as a substitute pharmacist, and saw to it that all of her children were educated and set on the path to successful lives. Then she left.

Upon receiving her divorce settlement, Theo decided that what she wanted to do more than anything was to return to school. She had spent countless hours over countless years studying the birds in her area of the western slope of Colorado and had been appointed by then Governor Dick Lamb to serve on the newly established Natural Areas Council. Worried about money and not wishing to relocate to eastern Colorado, Theo contacted Western State College in Gunnison, Colorado. She was encouraged to enroll there and worked from 1978 until 1982 to earn her MA in science in fresh water ecology. While there, she taught courses in natural resource conservation. Also while there, she pursued her desire to work at the Rocky Mountain Biological Laboratory. Her daughter had been a field assistant to Dr. Paul Ehrlich and Theo couldn't wait to meet him and other dedicated scientists at RMBL. Excited by this prospect and encouraged by a member of the faculty at Western State, she wrote to Prof. Stanley Dodson at the University of Wisconsin, who was to teach water ecology at RMBL that summer. Dodson encouraged her to attend the summer session in Gothic, and she did so as a student. At much the same time, her affiliation with the Natural Areas Council provided her with another connection to the place.

Theo's initial contact with RMBL had been through the Natural Areas Council. In 1978, prior to starting her graduate studies at Western State College, she was asked by the Council to create a power point presentation about water. The presentation was called, "The River; The Price of Development", and it followed the course of a river starting at its source and illustrating its uses. After giving the presentation to the then Undersecretary of the Interior, Guy Martin, she was invited to give it at RMBL. While there, she toured the Mexican Cut Preserve, which the Natural Areas Council was involved in setting aside for protection, and met Dr. Scottie Willey, who was particularly involved in protecting the Cut. She and Scottie "bonded", in Theo's words, and worked together, remaining friends after Theo's association with RMBL ended.

Theo recalls that when she arrived as a student in Gothic, she was housed with a more mature researcher rather than with the students. She was, at the time, fifty years old! This scientist had been an economist, turned entomologist and statistician. Theo credits her with helping her to get over her fear of statistics.

During her two summers in Gothic, Theo collected insects and took what courses were offered. She repeats a theme, heard over and over again, that the people at RMBL were supportive and encouraging, very much like a family. Since the purpose of a field station is to work in the field, collecting specimens for study in the laboratory – or establishing experiments *in situ* - she spent most of her days in the mountains surrounding Gothic and Crested Butte. She worked alone and with others, not infrequently paying the children of researchers to help with collecting insects. She recalls spending quite a bit of money to pay for insects brought to her by enterprising children. Through it all, she maintained a level of enthusiasm that is often seen only in the very young. Her own description of her work in the field reflects that.

> So my first collection spot [was] way up above – just about when

> you come down off of Kebler Pass and you come into the big open meadow. Maybe a mile down on the right was where I first started sampling. But then, also, I had to go over to the other side - the Oh Be Joyfuls; I went up Slate Road and climbed up – had to take horses to get in – and went up to the base in the back of the Oh Be Joyfuls. Spectacular! Oh my god! It was just a spiritual experience. And it just happened to be at the time of year when all the saxifrages were out – all these different kinds of saxifrages. If you want to see waxy, little, precious, beautiful flowers, that's the place to go.

Being afraid of heights tends to be a problem when one is working in the mountains, and Theo was acrophobic. However, she was able to complete her work with a little help from her friends:

> I'd get up there and look down and ... oh my god! All those acrophobic experiences! It was all so wonderful. Everybody helped, everybody was there! I couldn't be where I am today without them!

Theo Colburn left Gothic after two summers, but she credited it with being the place that helped set her on the course she pursued after completing her PhD in zoology at the University of Wisconsin-Madison. In addition to collecting insects at RMBL, she worked in the field with the team that would prove the impact of acid rain on the environment in the western part of the United States. Her post graduate work took her to various places around the country and she was responsible for discovering the link between chemicals released into the water and genetic distortions causing birth defects, known as "endocrine disruption". In 1996, she co-authored a book, ***Our Stolen Future: Are We Threatening our Intelligence, Fertility and Survival? A Scientific Detective Story.*** She also authored other books and scientific papers focused on endocrine disruptors and traveled the country and the world lecturing about the issue. Upon her retirement, she returned home to Paonia, Colorado, and founded TEDX: The Endocrine Disruption Exchange. TEDX is "an international nonprofit organization dedicated to compiling and disseminating the scientific evidence on the health and environmental problems caused by low-level exposure to chemicals that interfere with development and function, called **endocrine disruptors.**"[23]

Dr. Theo Colborn passed away on December 14th, 2014. To the end she continued to work actively with TEDX. She also joined the battle against harvesting natural gas from underground areas, known as "fracking". Until her death at age 87 she was a force to contend with and the commitment of those who worked with her will ensure that her legacy endures.

Women of science continue to be an integral part of the Rocky Mountain Biological Laboratory, and a significant presence in Gothic.

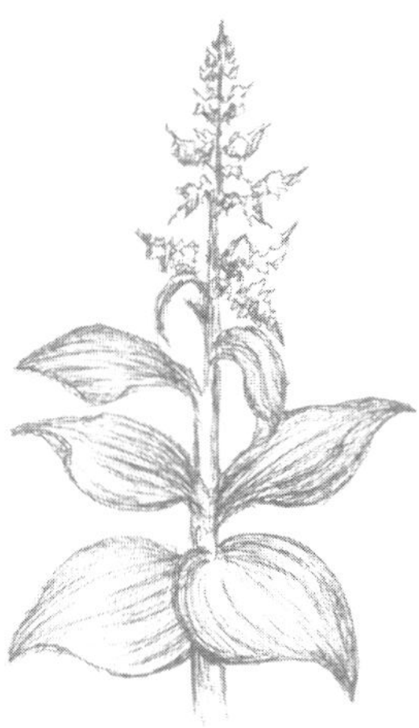

Jean Langenheim on Why Not to Use Pack Animals on Research Trips in the Mountains

Having spent multiple summers traversing the Colorado Rockies on foot with her husband, doing research for their doctoral dissertations, Jean Langenheim became somewhat of an expert on how to make those trips. In the summer of 1955, when working at the Rocky Mountain Biological Laboratory, she made several collection trips to remote areas with other colleagues. On one occasion, she was persuaded to use a horse as a pack animal. This is her story as to what happened and why she chose not to do it again. It is indicative of adventures at RMBL.[1]

The Prince of Beasts

On days that we weren't teaching [Bill] Weber [of the University of Colorado] often persuaded me to go collecting in remote areas. I not only soaked up the breadth of his botanical knowledge, but had several adventures with him as well. I especially remember a trip to Conundrum Pass on the Old Stage Coach route between Gothic and Aspen. This jaunt completely convinced me that Ralph and I had been right not to use pack animals for our dissertation research.

Weber and a Colorado University colleague who was joining us, Sam Shushan, were adamant about using a pack horse to carry our gear for the trip. I arranged for a horse from a Crested Butte tavern owner and with a couple of students went down to pick him up. It started to rain gently as we left, but by the time we got to Crested Butte, it was coming down in buckets. We struggled until late in the night walking the horse, named Prince, over the muddy roads in a driving rain from Crested Butte to a cowboy's stable near Gothic.

At 5:00 am I turned out of bed to find the sky absolutely clear. Shushan and I went to the stables and got Prince, and the packing procedure began. We took guidance from a book entitled Going Light with Backpack or Burro, which Ralph had purchased when we were planning a trip to the Wind River Mountains. Two hours later we had everything tied down and were on our way.

Prince, however, only wanted to eat. We literally had to pull him along. Somehow in the attempt to keep Prince going, we missed the turnoff to Conundrum Pass and arrived with much panting at Copper Lake. We staggered back down the steep grade and finally found the turnoff but then had to pull and tug that animal for five miles up a very steep, narrow trail across talus. We had not brought a canteen, as we incorrectly thought we would cross a number of snow patches. We were soon acutely dehydrated. Then, a third of the way up, the pack saddle began to fall off. The exertion of unpacking and repacking at circa 12,000 feet was terrific. We finally reached the pass at 6:00 pm and decided to go to Conundrum Hot Springs to spend the night. On the map, it looked like a short distance, but it turned out to be another two-and-a-half miles and a 3,000-foot drop in elevation. We finally made it around 8:30 pm. We cooked a steak over an open fire, devoured it with some tea, and crept into our sleeping bags.

Unfortunately, sleeping was not an option. We were up most of the time trying to chase away animals eating our food, and making sure that the horse was okay. Finally we got up at 4:00 am; we couldn't sleep, so we decided to get an early start collecting. We ate breakfast lethargically, then packed Prince. We were enormously proud of the job we had done. Now for the climb back up to the pass.

Prince of course wanted to eat every few feet, so we let him graze while we collected and took pictures. This went along fine for almost an hour, but then Prince suddenly took off galloping down the trail. We stood in amazement, watching as he disappeared from sight. Weber had even put our lunch on the horse, not wanting to carry anything extra. The three of us emptied our pockets; among us we had one chocolate bar, two ten-year-old malted milk tablets from a first aid kit, and a small package of lemon drops.

On the long trek back to Gothic it started to rain heavily and then came the final blow. We began to find evidence that Prince was progressively losing his load – we staggered back to camp carrying most of the equipment we found. Prince had been caught as he came through Gothic, but the next morning we had to go back up the trail to recover the remainder of our gear as well as the pack saddle. We found the panniers but the pack saddle had vanished; Prince somehow had gotten rid of it. We spent most of the afternoon trying to get arrangements straight with Prince's owner and returning him to Crested Butte. This adventure, except for a trip to Copper Lake two days later with my class, closed my field season for the summer.

Forward in Time — A Brief History

Life and science coexisted in Gothic at the fledgling Rocky Mountain Biological Laboratory throughout the 1930s and into the early 1940s. The success of the first five years created momentum that propelled the young organization forward, year in and year out, despite continuing financial concerns. The founders lived up to their commitment to underwrite the operating costs of the organization, most especially John Johnson and his family. Everything the Johnson family did, everything the Johnson children had during the academic years was the result of careful consideration for the needs of RMBL during the summer months. Faculty continued to be recruited to teach, students came to take classes, researchers arrived with their individual projects to pursue and the physical plant was maintained and improved. But money was always tight.

Not much changed in the way science was done throughout those years. The focus of the scientists who arrived was mainly observation and the recording of those observations. The label "naturalist" has not infrequently been applied to the scientist of the time. There was no laboratory equipment to facilitate the establishment of experimentation as we know it today, nor was that practice common. The mountains, the fields, the rivers and streams were the laboratories. One can imagine Harriet Barclay leaving her cabin each day and walking through the fields of wildflowers, choosing specimens to dry, identify and place in her herbarium. Afternoons or evenings might be spent cataloging the specimens and possibly painting them to preserve for future botanists the true colors of the flowers. There was no need to go afar – Nature was right outside each cabin door.

As with most any organization, balancing the books was a top priority. The treasurer's report for fiscal year 1938-1939, ending July 5, 1939, reveals a meager balance of $281.83 with which to begin the 1939 summer season. This was an increase of less than five dollars over what the beginning balance had been a year earlier. The surplus for the upcoming year was estimated to be only about $58.00. Expenses were low, mostly consisting of payment on debt, insurance and printing costs. Income would be from tuition, cabin rental and dues – but this would all be offset by operating expenses including meals, gas and supplies. The bottom line was that RMBL was just getting by. Members were often negligent about paying their annual dues, to the point where the treasurer, Dr. George Hunter from Wesleyan University, recommended to the Board in 1939 that fourteen members be dropped from the rolls for failure to fulfill their pledges.[1]

The change of decade to the 1940s coincided with conflict in the world. Many young men were off to Europe or the Pacific to fight in World War II. Science was forced to take a back seat to politics and RMBL was only one of many organizations impacted by the cataclysm. The Director's report dated January 23, 1947 informed the membership that, "Due to the war, most of the buildings had not been used since 1941, so numerous minor and major repairs had to be made, including new roofs for the kitchen and two student cabins."[2]

Nineteen-forty-six, however, found people returning and Gothic being revitalized. Two additional student cabins had been completed just prior to the opening of the season, allowing for the housing of up to eight more single students or two married couples. The session was judged to have been quite successful, attended by "fine quality university and college students coming from several sections of the United States… the majority [of whom] were graduate students working toward the Masters and Doctors degrees".[3] The food situation was challenging, due to limited availability and escalating costs, post-war, but it seems everyone was adequately fed. And the session ended with yet another very small surplus. It was hoped that butane gas and some electricity might be installed, but that would depend partly on the positive response of members in the timely paying of their dues.[4]

August 2nd, 1952 was the occasion of the 25th Anniversary Celebration of the Rocky Mountain Biological Laboratory. Activities were planned throughout the day, with a mountain trout picnic anchoring the noon meal. A letter from Director John C. Johnson informed interested parties of the plans, as well as summarized the successes of the twenty-five years since the inception of RMBL in 1928. He proudly recounted that, "high academic standards have been maintained," every instructor had boasted a PhD, credits earned at RMBL were accepted in universities throughout the United States, "so far as we know". The physical plant had grown, increased by "a large log laboratory, a dining hall, a kitchen, seven student and three faculty cabins". Four faculty members had leased sites from RMBL and built their own cabins, and four of the buildings remaining from mining days in Gothic had been "improved for living quarters for students and one for a laboratory". In addition, a shower house had been built and electricity had been provided for the two laboratories, the dining hall, the kitchen and the Director's residence. More, of course, was needed – specifically a "small Library Building and study room, about twenty by thirty feet in size", as well as books and journals to stock the shelves.[5]

Much of the work over the years was done by the students and faculty during the summer months. Money was too tight to hire general contractors, or even many day laborers. The tradition of community cooperative support through manual labor had become engrained in all summer residents of Gothic. Indeed, in 1952 Dr. Johnson reported that an instructor, Dr. W.H. Emig, Professor of Botany at the University of Pittsburgh, had studied the art of masonry "for fifteen hours a week, all year" in order to be educated in the trade so he could direct the building of the new library. The people he directed were the faculty and students, undoubtedly donating "many hours of free labor".[6] Such cost cutting measures and cooperation in these early years of the 1950s made it possible for John C. Johnson to finally be fully reimbursed by the organization for his personal expenditures over the years on its behalf. The time volunteered by the Johnson family, the other founders and all the people who had been a part of the growing organization could never be evaluated, let along compensated.

The mid-1950s marked the beginning of major upheaval occurring in the heretofore enchanted, frontier existence in Gothic. The challenge did not come from the everlasting nature of the environment, but from changes in the field of science itself. Observing and recording were becoming the way of the past. Hypotheses and experimentation were the wave of the future. The scientists still came, but the students who came with them were of a new generation. They had fought a war and been successful. Part of the success of that war effort was the result of technology stemming from new scientific discoveries. Biologists might not be technological innovators - yet - but the processes of discovery extended throughout all fields of science. And the government had cash available and the will to spend it.

Enter the **National Science Foundation.**

In its own words,

> The American people recognized that scientists and engineers had helped win World War II. Penicillin and the atomic bomb were but the two best known of the many contributions made by the research community. With the coming of peace, the challenge facing politicians and researchers alike was how to ensure that science and engineering would continue both to expand the frontiers of knowledge and serve the American people. The answer was the

> National Science Foundation (NSF), established in 1950, which continues to be the only federal agency dedicated to the support of fundamental research and education in all scientific and engineering disciplines. Charged with making certain that the United States maintains leadership in scientific discovery and the development of new technologies, the NSF has provided funding for thousands of distinguished scientists and engineers to conduct groundbreaking research, including many Nobel Prize Winners.[7]

Created by Congress in 1950 as an independent federal agency, the National Science Foundation took its mandate "to promote the progress of science; to advance the national health, prosperity, and welfare; to secure the national defense", seriously. From an initial appropriation of $225,000 in 1951, the budget of the NSF grew to 7.2 billion dollars in 2014. It is responsible for approximately 24 percent of all federally funded basic research in America's colleges and universities.[8]

It is not uncommon that when money is involved, disagreements are sparked, multiply and escalate. Money also often translates into power. This is the very foundation of what would become known among the members of the Rocky Mountain Biological Laboratory as the **Gothic Wars.**

For almost three decades, the individuals who sustained the Rocky Mountain Biological Laboratory came from colleges and universities located mostly in the heartland of the United States. Faculty members at Ivy League universities of the eastern part of the country were more likely to be affiliated with research stations elsewhere, if at all. This began to change in the early 1950s. Partly due to the need for money, partly to the desire for prestige, partly to survival instincts among the entrenched founders and supporters of RMBL, the leaders of the organization, especially its Director and Founder, Dr. John C. Johnson, came to believe that growth throughout the second half of the twentieth century would best be fostered by recruiting scientists from elite schools to work in Gothic. Ideally, big names would bring money and money would allow for growth. The individuals who were actively involved in the difficult times that began as a result of this process are no longer around, and the surviving documentation to support the facts is sparse, but a combination of those papers and memories of people who came later and knew the story has helped to rebuild the following tale.

Dr. Johnson had been Director of the Lab since its inception. He was a science professor during the academic year, but not in reality a research scientist. His true calling was that of administrator and protector of the interests of RMBL. At some point in the mid-1950s he is said to have realized that future growth required the infusion of new, young scientists who were active researchers in their fields. These individuals were dubbed "geneticists" by the old-timers at RMBL, as they were studying chromosome rearrangements, genetic adaptations to environments, physiological and behavioral aspects of adaptation in insects such as butterflies and in plants. In Dr. Johnson's words, "I believe they are trying to find out what really happens to the chromosomes when things change; and to see what influence, if any, is caused by the environment".[9] This was indeed a change in course for the researchers at the field station, who had mainly focused on observation of the flora and fauna outside their doors. And, this change would require an infusion of cash. Dr. Johnson recorded in 1955 that, "Perhaps a thousand dollars has been used to build three temperature control rooms inside the large log Ecology Research Laboratory and three other research cubicles".[10]

In ways that may not be specifically known, perhaps at conferences or through the grapevine of academia, Johnson was successful in recruiting new, young scientists to Gothic. They came from Yale and Princeton and Harvard in the east, as well as from Missouri and Wisconsin and other schools in the middle of the country. Colorado and California, Ohio and Pennsylvania were not left out of the new line-up. They brought students, and they brought grant money. In 1955, the researchers at RMBL and the station itself were recipients of five National Science Foundation grants. Four of these grants went to scientists; one, for six thousand dollars, went directly to RMBL for the purpose of financing the installation of electrical power and a water line at the laboratory.

As money flowed in, as much or more flowed out. Whatever system was being used for accounting was found, by some, to be less than adequate. In January 1957, a letter from Professor Charles Remington at Yale University to the Executive Committee of RMBL focused first on disorganized land ownership and administrative issues. He detailed problems with leases that highlighted no chain of ownership; leased land that did not belong to RMBL and was, therefore, not theirs to lease; plots that were unreasonably large; and expired leases that had been extended in opposition to RMBL limitations. He stated that, "the existing leases are completely chaotic" and must be rewritten. He then went on to say, "THE STATE OF FINANCES MUST BE CLEARED… I was shocked to find that several bills I sent in for payment sat around for weeks and some may still be unpaid, as far as I know… I have just submitted a large grant request to NSF and decided to send it through Yale, not RMBL, in part because the administration of funds has been so haphazard".[11]

Remington was not the only one to voice concern about the state of finances at the Lab. Financial papers in the RMBL archives in Gothic indicate there were insufficient funds to pay the bills in 1956 and 1957. In an October 1956 communication from Executive Committee Treasurer, Dr. J. Teague Self, to the President of the Committee, Dr. J. Fisher Stanfield, Self complained, "Not only does RMBL not have money to pay for an audit, it does not have the funds to pay the electric bill in January... I am rushing this letter to you to announce that the honeymoon is over and that it is time for the boys to cease the splurge."[12]

There was also concern about the integrity of the RMBL facility and the lands on which it was situated. Dr. Herb Levi had been spending summers in Gothic since a graduate student in 1947. He continued his association with RMBL as a professor of zoology at the University of Wisconsin and then professor and Curator at the Harvard Museum of Comparative Zoology, until about 1968. He and his wife built a cabin that continues to bear his name. In 1957 he wrote to Executive Committee member, Dr. Colin Pittendrigh of Princeton University, about the damages being inflicted upon areas of scientific study and experimentation. The children and pets of scientists were running around, unsupervised, and trampling plants; flowers were being dug up; insects were being destroyed. A "visitor" had to be talked out of pulling out mistletoe by the Barclay cabin; he claimed to dislike it because "it smelled". Areas near the creeks had been dug up for fishing worms and were likely to wash out in heavy rains. Levi beseeched Pittendrigh to make it understood to one and all that "exotics not be brought in, pets be leashed, and as little disturbance as possible be done to natural vegetation and fauna".[13]

Levi blamed the arrival of "more and more people (with their children) whose only contact with and appreciation for nature has been through a microscope tube or in the laboratory. These people need to understand the problems of a field researcher". It seems the lines were being drawn between the observational biologists and the "geneticists".[14]

Levi then turned to the subject of money, and how to obtain endowments for the financially struggling organization. He suggested that individuals who might be willing to donate money would naturally be interested in being a part of the organization. Opening up membership in RMBL to such persons could be a great incentive to them. He was concerned that a cliquish attitude among the station biologists would prevent them from reaping financial rewards. He went on to say, "Do you know of one other scientific society that is as snobbish as the R.M.B.L? What society would reject a member who is sincerely interested? Even some professional societies may be easier to join than the R.M.B.L."[15]

All of this is by way of setting the stage for internal conflict and explosion.

In response to an apparent, unspecified attempt to change the constitution of the organization in late 1957, Dr. J. Fisher Stanfield, a RMBL Trustee, wrote to Chris Johnson, son of the founder, that the constitution should be left alone. He had written to Charles Remington asking that the vote be forgotten or covered up, "to save face". "If he [unspecified] refuses, then we shall be forced to bring it out which may well result in a schism and we are near enough to that as it is. He has assumed authority in several ways that cannot be substantiated by the written records. Let us charge it to impatience or youthful outlook." He went on to say, "The boys in the cloistered field of genetics consider themselves as quite the top stratum professionally, but I hardly think the record supports that view." Stanfield was willing to overlook much in order to keep the peace, but ended by saying, "I have no intention of sitting and taking lip from anyone since I do not have to do so."[16]

What happened during the summer and fall months of 1958 is a testament to the damage that can be wrought through lack of cooperation. The adjective "childish" has been applied by many who know of the events and the behavior of the scientists concerned. The damage was severe, both morally and financially, and RMBL quite possibly could have seen its final seasons. Cooler heads prevailed in the end, but the lesson and the scars remain in the collective memory of the organization.

President of the Board of Trustees Dr. Colin Pittendrigh, had requested a conference with Director John C. Johnson in June of 1958. They met at Pennsylvania State College and Pitt (as he was generally known) told Johnson that it was time he relinquished the directorship of RMBL, a position he had held for thirty years. Johnson was understandably reluctant to surrender responsibility for the organization he had birthed and nurtured. Pitt's reasoning, at least initially, had to do with finances. He was in the process of applying for a $100,000 grant for RMBL from the National Science Foundation and argued that NSF would view the request more favorably if the person who ran the organization was a research scientist from a top tier university. Johnson agreed to think it over and discuss it with his wife, Mildred, and son, Chris. A week later, he wrote to Pitt and said, "At present we think there is considerable merit in the tentative conclusions you and I arrived at, so far as my personal relationship to the R.M.B.L. is concerned". He went on to say that before coming to any final decision, the "present and future polices and plans of the R.M.B.L." should be reviewed at the annual meetings of the Corporation and the Board of Trustees.[17]

Professing his continued sense of responsibility to RMBL, Johnson conceded that, "If being Director is where I serve best then I should remain as Director, but if I can serve best as President of the Corporation and Board of Trustees, then I should be willing to make the change".[18] In a later letter sent to the Trustees and Officers of RMBL, he informed them that he would be "quite willing to retire from my present long-held position as Director. Indeed, I am anxious to relinquish the job, but would not voluntarily do so until I were sure that the position would be adequately filled".[19]

Herein lay two issues that would become explosive. One was Johnson's insistence that he remain involved as the President of the Corporation and Board of Trustees. The other was, quite simply, who would be acceptable as a replacement for him as Director?

One can only speculate as to the sense of concern and inevitable political maneuvering that followed these events. Subsequent to Johnson's announcement that he would be willing to leave pending agreement on an "adequate" successor, a meeting of the Board of Trustees was held at which there was a bare minimum quorum present. Harriet Barclay of the University of Tulsa nominated a colleague of hers from the same school, Dr. Ralph W. Kelting, to replace Johnson. Kelting was visiting Gothic at the time and was not even a member of RMBL. He was persuaded to accept the nomination with the understanding that the president and dean at his university must allow him to spend summers at Gothic, that Chris Johnson (the heir apparent) would assist him during his first summer and, if it should not work out, that Chris Johnson agree to take over as Acting Director, something Chris did not wish to do. A massive outcry from the membership objecting to the choice of an "outsider" was to no avail. The vote was taken by the Trustees with the bare minimum quorum in attendance and Kelting was chosen as Director. To add insult to injury, there was discussion about extending the powers of the Director to include voting power on most committees and administrative control over teaching and research. The course was set for a major confrontation.[20]

There was so much to be angry about, on both sides. The old timers felt strongly that Dr. Johnson was being railroaded out of office and herded into retirement. By holding onto the Presidency, which he did at this time, he was at least able to keep himself involved with the organization he loved. But the younger generation was furious that someone with no experience would be placed in the Director's slot, and objected to Johnson staying on in any capacity.[21]

Angry letters flew back and forth among the Trustees and Johnson. In one, it was pointed out to Johnson that "your accession to the Presidency of the Board could not be made a <u>condition</u> of your retirement as Director, no matter how much we felt personally that such a move would be suitable".[22] Another argues that, "if the Trustees assembled this summer have so little foresight as to appoint a man as Director who has never been associated previously with RMBL, who has had nothing to do with the effort to obtain the $100,000 grant from the National Science Foundation and, I'll bet, who has never administered a sizable government grant before, then only a fool would vote to extend further powers to such a director."[23]

Johnson tried to diffuse the anger. He placed a phone call to Pittendrigh in mid-September that was an unmitigated failure, so he followed up with letters. He claimed that he had welcomed Pitt's request that he resign as Director, as he had "important reasons, personal and otherwise", to leave office at that time. He also defended the choice of Kelting as Director, reasoning that "it is important… for a Director at Gothic to understand the psychology of the cattlemen, the western businessmen, and get along with their organizations and this is more important than you realize. The Trustees know Dr. Kelting to be an able biologist, an executive, a good public relations man, and felt extremely fortunate to have procured his services". He did not address the issue of himself staying on as President.[24]

Colin Pittendrigh was so incensed by the actions of the Board and John Johnson himself that he resigned from the Board of Trustees, of which he was President, resigned his membership in the Rocky Mountain Biological Laboratory and withdrew his name from RMBL's grant application for $100,000 from the National Science Foundation. His stated reasons for his actions were, firstly, that Johnson had made public their private conversation about Johnson stepping down. But he then expounded,

> "We have already received approximately $10,000 from NSF on the premise that the station was to develop as a summer center for studies in the New Systematics, as a center that is for all the disciplines like taxonomy, ecology and genetics that contribute to present day evolutionary theory. And our pending application for a further $100,000 is also based on this view. It has been clear then for some time that we needed a director familiar with these subjects in their modern form and familiar with the people we have **begun** to attract to the laboratory in support of such a program… [M]y clearly stated intention to withdraw from the NSF application if you retained the Directorship was not a threat intended to further a mere personal hope for your retirement; it was based on a carefully formed judgment and a considered view of my responsibility not only to the station but the profession as a whole including the National Science

Foundation.[25]

Pitt had also expressed his surprise and disappointment at the choice of Director in an earlier letter to Johnson. "I have no reason to doubt the judgment of all of you who elected him but it had been my impression that we were all to search for a suitable persona and consult with each other before taking any serious step… Also, I feel that with the trustees meeting limited to the barest quorum this was indeed a major step to take. What I have in mind here, of course, is that many of the more able [*sic*] members of the Board of Trustees were among those unable to be present this year and, therefore, excluded from this decision".[26]

Tempers continued to run high. In October 1958, Professor William K. Baker from the University of Chicago wrote that he was deferring his plans to build a cabin the following summer. "If RMBL is to be just a pleasant place where people allied, however faintly, with biology can get together in good fellowship with some of their expenses paid if they give a course, then I am not interested. From a career viewpoint, I cannot afford to spend two months of the year on this type of a junket. What I am looking for is an intellectually stimulating environment away from the hubbub and minutia of a university; a real association of productive scholars. This is the type of association which I wish my students to experience."[27]

Demands were made for the resignation of both Kelting and Johnson. Concerns were raised over the future of grants from the National Science Foundation. Suggestions were put forth for who might be an acceptable Director. Moral support was offered to Johnson, but not without a caveat. John Pelton wrote, "In spite of their insults… I feel that we should try and keep people like [Edward] Novitski, [William] Baker and especially [Alfred] Sturtevant with the RMBL for the sake of the survival and growth of the Lab itself. I don't see how RMBL will survive financially otherwise, without some outside support which the above group can materially help obtain. I'm sure that their 'taking over the lab' can be controlled by others of different interests than theirs, and anyway, that we must take the chance."[28] Survival of the Rocky Mountain Biological Laboratory was paramount.

Reaction to the resignation of Pittendrigh was received with anger and with relief, depending on the side of the conflict one was on. In a letter from Chris Johnson to his father, he encouraged him to not blame himself for what had happened. He said that his initial reaction to the resignation was that it would not be beneficial to the Lab, but he did not believe that NSF would kill the grant. He expressed concern as to whether or not some of the scientists would abandon the Lab, but supported his father wholeheartedly. He predicted that, "In the long run we may be better off to be rid of Pitt; he had turned out to be thoroughly unreliable and full of self-ambitions". In a moment of familial loyalty, and uncharacteristically for Chris, he told his father, "I personally would like to sock Pitt in the nose for his insults to you especially, and for his nasty remarks about the Lab in general. But of course I'll refrain from any such tactics".[29] He also, curiously, reminded his father not to let anyone know that he, Chris, was applying for a grant from NSF.

One trustee, Dr. Reginald Manwell, wrote to the National Science Foundation expressing his regret for Pittendrigh's action, "which seems to me was taken in a spirit of personal pique," and proffered his confidence in Dr. Johnson as President, and in the remaining members of the Board of Trustees.[30] But only two days later, Novitski wrote to the Trustees that, "The letter from Pittendrigh was the most serious blow to the futures of the RMBL that could possibly have happened – yet it is clear from his letter that he had no choice... <u>On one point there must be no misunderstanding.</u> The appointment of a virtually unknown biologist as Director, and its inevitable consequence, the loss of Pittendrigh as a member of the Corporation and Trustees, has decreased our chances of getting funds from NSF from whatever they were previously to precisely <u>zero</u>."[31]

So how did all end? Dr. Kelting resigned in October. The National Science Foundation denied the grant, "not entirely unexpected in view of the disgraceful August events".[32] A compromise candidate was found for the position of Director. On December 31, 1958, Dr. Johnson informed the Board of Trustees and Officers of the Lab that Dr. Robert K. Enders from Swarthmore College had been elected as Director of RMBL, and Chris Johnson as Administrative Assistant to the Director. The letter was signed, "John C. Johnson, Chairman of the Executive Committee and President of the Corporation of the Rocky Mountain Biological Laboratory at Gothic."[33]

The battle was over, but the war continued and the wounded remained.

The 1960s found yet another new generation of young scientists arriving in Gothic. Students of the entrenched researchers, they cut their field station teeth at RMBL and some have stayed on into the twenty-first century. They were not ignorant of the difficulties and some were even marked by them but, in the words of one established scientist, "as a result of some of the turmoil that was going on at the time, I made it my business to really try to find out what the origins of the turmoil were, to see whether it was worthwhile continuing my work here; at least post-Ph.D. work.

[Other people] also had to make that decision. Some of us did hang tough and I think it's good we did."[34]

The early years of the sixties continued to be fraught with dissention among the ranks and animosity toward the administration of RMBL, specifically Dr. Enders. An open letter to the membership in June of 1963 written by Dr. W.K. Baker and distributed as the first annual Newsletter in years, explains the situation in detail.[35] The letter begins with an apology for the long duration between Newsletters and explains that "RMBL is having difficult times and we have hesitated about passing this information on to the members… Facts, rather than hearsay, may clarify our difficulties".[36]

The National Science Foundation had created a program called the Undergraduate Research Participation Program (URP), the goal of which was to include undergraduate students in research activities of practicing scientists. The Rocky Mountain Biological Laboratory had received two grants from NSF between 1960 and 1962 to participate in this program and had done so successfully. It was considered to be a boon to both the education of young college students and the financial coffers of RMBL. Dr. Edward Novitski was the individual credited with encouraging RMBL to apply and taking responsibility for preparing the initial grant requests; he was also a member of the NSF review panel for the program. By 1962, however, dissatisfaction was being voiced by some in the membership over the administration of these grants.

Dissatisfaction centered on several issues. It appeared that the Director, Dr. Enders, had awarded several students participation grants in the program and permitted them to arrive in Gothic without informing anyone, least of all the "investigator", or research scientist, under whom their research was to be conducted. As a result, planning for their participation and providing for their equipment needs was impossible – and the investigators arrived at RMBL with no idea they were mentoring undergraduates. To add fuel to the fire, Dr. Enders demanded at the meeting of the Corporation in 1962 that he, the Director, have sole authority to select the students who would participate and, unless this was agreed to, RMBL "would have no part of any URP program".[37]

Dr. Levi, in a rebuttal to Dr. Baker's version of the story so far, said, "Everyone realizes that to add the burden of running the RMBL to other tasks, especially at the end of the school year, makes shortcomings inevitable. That the director has occasionally done a poor job in putting students in contact with investigators before arrival is an unfortunate result of late admissions". He also stated that the information in Dr. Baker's newsletter was unauthorized and unverified by "any officer of the corporation".[38]

More seriously, this already tenuous situation was complicated by the fact that Dr. Enders and Chris Johnson had prepared and submitted a new URP grant in the fall of 1962. They took the liberty of listing on the proposal names of investigators who would be working with the program – but did not inform those individuals of their inclusion in the application. "Contrary to the procedure that was rigidly adhered to in the past, the writers of this application listed in it the names of the investigators, the number of undergraduate research participants they would sponsor each summer, and a description of the manner in which their individual programs would be run, without obtaining the permission of the investigators or letting them know that their names were being used to obtain a grant from a federal agency." This would have possibly been acceptable had the investigators names been listed only as "past participants" of the grant program, but they were not. They were included as "active supporters", something which turned out to be patently untrue.[39]

Again, Dr. Levi disputed this version of events, claiming that "it was taken for granted that those associated with the laboratory would help and their names were listed… Drs. Baker and Ehrlich protested and asked that their names be removed from the application because they had not been asked for specific permission. By carrying his protest directly to the NSF, Dr. Baker for the second time jeopardized an RMBL grant to the NSF."[40]

Dr. Novtiski, in a lengthy letter to the RMBL Membership in response to Dr. Levi's letter, stated that he had been asked by NSF to serve on a panel evaluating URP applications. "Quite to my surprise, one of these was submitted by RMBL." He noted that there were "four or five persons" listed as supporting the application whom he knew "from conversations with them during the preceding summer had serious objections to the program at RMBL." He stated, that "for training applications, where the merit of the application must depend primarily on the qualification of those who are to do the training", it is vital that those individuals be aware of participation in the application. He debated what to do and decided to inform the panel of what he knew.[41]

Novitski went on to explain that the panel did not recommend the application be rejected. "Instead, it was returned to NSF with a request that NSF officials look into the possibility that it was written up in a completely misleading fashion, with those officials then to take appropriate action, depending upon their findings. Dr. Enders was given the opportunity by NSF to remove the questionable sections and to resubmit the application. This, he refused to do. NSF then rejected the RMBL proposal."[42]

Whatever the source, the National Science Foundation learned of the controversy and did not approve the grant. According to Baker, Chris Johnson apologized "courteously" to those persons whose names were used without permission. According to Levi, the "Executive Committee of the RMBL" considered removing members from committees who were "unable to cooperate". However, no action was taken because "Drs. Baker and Ehrlich apologized." Whoever did or did not apologize, the confrontation was not ended and the situation was not diffused. If anything, it became even more convoluted.

After NSF had made its determination, Dr. Baker said he wrote to them and asked that his name be removed from the application. He and Paul Ehrlich wrote to the President of the Board of Trustees, Dr. Manwell, objecting to the way the application had been drawn up. Dr. Enders insisted that Dr. Ehrlich was not listed among the investigators, but Dr. Novitski had a copy of the application sent to all members of the Board. It was obvious that Dr. Enders had not been telling the truth. The application contained a paragraph about each of the named investigators, including Dr. Ehrlich, describing what their future participation would be. "An investigator participating in a program would be considered by any granting agency as a supporter of that program"; therefore, each of the scientists named had been included in the grant as supporters of the program without their consent.[43]

To still further complicate the situation, Dr. Enders had been charging the students enrolled in the URP program one hundred dollars each from their personal stipends to pay for tuition and/or laboratory space. This was in direct violation of NSF rules, especially since RMBL had received separate monies to pay for the research expenses of each participant under the grant. Although it is possible this may have been a misunderstanding on the part of Enders and the administrators, a letter from the Program Director of NSF Undergraduate Science Education deemed it "…a definite deviation from administrative conditions outlined in the letter of understanding accompanying all Undergraduate Science Education Grants". The Board agreed to take under advisement a way to rectify the mistake. Novitski insisted that the students who had been unjustly charged the monies be refunded by the Board, with six percent interest. This was never done, as the NSF instruction merely noted that, "In view of the time which has elapsed, we feel that the question of return of student fees is a matter for determination by the RMBL board".[44] Novitski was outraged, saying "It probably never occurred to the NSF officials that the board of trustees would say no to such a clear moral obligation!"[45]

Another URP grant request was submitted to NSF in the winter of 1962; it did not include the names of those investigators who objected to being included. This grant was also denied. Novitski claimed that he told RMBL that the grant was turned down due to "lack of merit".[46]

Dr. Novitski, in his role as a member of the NSF review panel and who had enlightened the RMBL scientists about the facts of the URP grant application, was asked to resign from the Board of Trustees, to resign his membership in the Corporation and to relinquish the lease on his cabin. The stated reason for this request was that Novitski had made statements in "undocumented" personal letters about the two men and as such was no longer welcome in Gothic".[47] In a letter from Dr. Herbert Levi, Secretary for the Officers and Executive Committee of the RMBL Board of Trustees, he stated that, "During the week of 18 May [1963] the Trustees took unprecedented action and voted formally (by a two-thirds majority) to request Dr. Novitski's resignation". He was to receive "fair compensation for his lease and cabin".[48] Novitski complied with their wishes and continued his successful career as a geneticist without benefit of summers in Gothic.[49]

In the "Newsletter", Dr. Baker prolonged his recitation of details of the story with a bit of editorializing. He lamented the rift in the ranks of the membership and feared for the future of the organization, identifying the problem as "a breakdown of communications between, and a breakdown of respect for, others whose sincere opinions differ from our own". He cited a poem of unknown authorship that he had recently read as summarizing the situation:

There's danger for people of all nations
In a lack of basic communications
And it's possible to go under
To the roar of self-made thunder[50]

Creating and identifying problems is one thing; solving them something entirely different. Pointing out that RMBL was a loose confederation of independent scientists who had "no legal or financial obligation to the Director", Baker conceded that the ties people had to the Lab "include our allegiance to what RMBL has done for biology in the past and its potentialities for the future." He suggested that in order to foster the necessary democratic basis of the organization, weekly meetings of administrators, teachers and researchers should be scheduled. To his chagrin, this had never been done by Dr. Enders.

How were the problems to be solved? "It has now become obvious that an outside judge of our past biological accomplishments and our future prospects, namely the National Science Foundation, has lost confidence in the ability of the present administration of RMBL to carry out meaningful programs… Unless the present administration revises its modus operandi, or unless we get a new administration, there is just no future for RMBL as a biological station that is in any way commensurate with its potentialities."[51] Dr. Levi also conceded in his letter to membership that, "The NSF has made it clear that RMBL will receive no more funds from that source until we are able to settle our internal disputes."[52] Sadly for the organization, Dr. Enders had lost much support among some of the membership, support which would be difficult to regain. And, he had no intention of resigning. Levi concluded his letter by saying, "if the RMBL is to survive as an independent, unaffiliated field station for biologists, we will have to continue to depend upon constructive efforts of all members and trustees."[53]

Yet one more potential crisis was averted in 1965. Correspondence among members of the committee appointed to nominate individuals for two expiring Board of Trustees positions reveals a conflict concerning the continuing role of John C. Johnson. The controversy erupted after Chairman of the committee, Dr. J. Teague Self, wrote to Johnson on behalf of the committee members to inquire if he wished to continue on the Board "now that John, Jr. is also a Trustee".[54] A few weeks later, Self received a letter from Chris Johnson, the "John Jr." in question, commenting on "correspondence between you and my father concerning possible nominees for RMBL trusteeship".[55] Chris was bristling with indignation as to the suggestion his father should leave the Board.

Ever one to jump to his father's defense, Chris wrote that, "If there be question whether two Johnsons should serve simultaneously on the Board, the decision clearly concerns whether I should be nominated when my father is already a trustee, not the reverse." Chris had been appointed several months earlier to replace Charles Remington, who left his position before its term expired, and he strongly disagreed with the suggestion that because of his new role, his father should not stand for reelection. "This is like saying that when a five horsepower motor is turned on the much needed one hundred h.p. one should be switched off!" Apparently there had been some discussion about creating an honorary trusteeship for Dr. Johnson. Chris agreed that the idea had "considerable merit, but I don't think it necessary now and perhaps it would at this time lead to misunderstanding." He indicated he would be willing to resign if asked to do so, but added that "this is perhaps not an appropriate procedure". He concluded his letter by saying, "I do think my father deserves distinct honor".[56]

Dr. Self immediately conveyed this information, along with copies of his correspondence with Dr. Johnson, to Dr. John Pelton and Dr. David Cameron, his fellow committee members, saying that he had written to Dr. Johnson "while under the delusion that he was ready to step down… Obviously, this is not the case."[57] Continuing with his reasons for pursuing this course of action, he wrote, "Confidentially, between us… there are some important members [of the RMBL Corporation] who have told me that they feel it is time for Dr. Johnson to step down… If he had volunteered to step aside and let Chris represent the Johnson name on the Board, everything would have been easy". However, Self had no desire to create a new conflict within the Board, so he suggested one might be avoided by resubmitting Johnson's name.[58]

John Pelton agreed with this suggestion, but David Cameron did not. In response to Self's letter, he wrote, "If the board of trustees has some important decisions to make and are not to be considered rubber stamps for somebody, then I should recommend a younger man".[59] He labeled the nomination "purely political" and said, "How can one argue with a man seventy years old and as inflexible as the senior Dr. Johnson? Must one put him in the driver's seat in order to expect any services from him? At what price are we to avoid painful decisions now only to drop the question into the lap of a volatile, oxygen starved, summer meeting?" He went on to express concern that the "men on high" among the trustees might not be able to "speak with authority to NSF or any other body" and to suggest that it would be a good idea to put someone on the Board who would "really enter into the modern spirit of the laboratory. Give Dr. Johnson honorary whatever you wish and what will sooth the wounds that may arise".[60]

A stalemate having been reached, Self conveyed this information and the reasons for it to Chairman of the RMBL Board, Dr. Reginald Manwell. There was insufficient time to revise the by-laws to create an honorary trusteeship, as had been discussed, and there were "obvious" mixed feelings among the trustees as to Johnson's re-nomination.[61] Therefore, a compromise was reached. Rather than simply nominate two persons for the two vacant slots, the committee decided to nominate two for *each* slot. Dr. Self pointed out what he called "paramount" issues to his committee members. One was that Dr. Johnson should be re-nominated, "particularly because of services which he is in a position to render relative to the tax status of the RMBL". The other was that Dr. George W. Hunter, whose term was also expiring, should be re-nominated because he had served only part of a term and he had "demonstrated a willingness to work objectively for the welfare of the RMBL". The next step was to decide which names should be placed in nomination along with these two.[62]

Several names were put forward and the decision was made. The

Nominating Committee conveyed its recommendations to the RMBL Board in June.

1. Replacement for Dr. John C. Johnson
   a. Dr. John C. Johnson
   b. Dr. Kenneth B. Armitage
2. To replace Dr. George W. Hunter III
   a. Dr. George W. Hunter III
   b. Dr. Ferchau, whose initials and address are not available to me[63]

A potential unpleasant situation had been avoided by compromise. Dr. Johnson was, indeed, re-elected, as was Dr. Hunter. Johnson served another four years and Dr. Armitage was elected in 1969.

When people become entrenched in their positions, it is difficult for those positions to change quickly. Time helps. So does agreement on small things that can lead to agreement on bigger things. Perhaps the most important part of this story is the *legacy*. Robert Enders, a compromise candidate, would serve RMBL as Director until 1967, when he was succeeded for a single year by Dr. Keith Justice. In 1968 Dr. John C. Johnson Jr., "Chris", was elected Director. Chris Johnson is the person credited with keeping the peace and holding the Lab together through the turbulent transition years from 1958 forward. He is fondly and universally remembered by those who knew him, and those who knew of him, as a calm, quiet presence who never took sides, never said a bad word about anyone. He listened but didn't tell tales. Not even to his wife, Dorothy, who was at his side through it all. Chris remained as Director until 1977, working to rebuild trust and maintain balance. He was, in the minds of many, the savior of the Rocky Mountain Biological Lab and its shepherd through difficult times.

Consolidation and growth became the trending factors surrounding the field station in Gothic, even as tempers cooled and new scientists with no history of involvement in the "Gothic Wars" arrived. Those scientists who did stay on did so with a sense of commitment, which continues on through time. They brought with them not only students, but their own research grants, which helped to maintain the purpose of the organization. In the words of Chris Johnson., the primary role of the Director was, and continued to be, to "promote the quality of instructional and research programs".[64]

The passing of the torch of the Johnson family saw the Directorship awarded to Dr. Francis Hunter in 1978. Dr. Hunter became quite ill during that year and died. He was followed in office by Dr. Peter Broussard for a three year term. Dr. Broussard's tenure in office "marked the beginning of an explosion of additional successful grant proposals." He was assisted by a professional fund-raiser, Marilyn Shaw, as well as by many volunteers in the Gothic community. New scientists were successfully courted and facilities were improved. A new lab was built, which would eventually be named "The Robert B. and Ruth L. Willey Research Laboratory".[65]

Infusions of capital and personnel required more constant attention on the part of the Director, especially, and the Board of Trustees. Broussard's successor, Dr. Andrew Beattie, was to be the last part-time Director. He served a three-year term ending in 1984. Up to this time, all Directors had served in that position, in addition to fulfilling their teaching and research commitments at their home colleges and universities. Beattie convinced the Board of Trustees that this had become cumbersome to the point of being inefficient and RMBL's first full time Director, Dennis Johns, was hired in 1985. Johns, like Hunter, was destined to serve only one year in office. An external search was conducted and Susan Allen, an ornithologist with years of field experience in the jungles of South America, was hired. Allen (who eventually married in Crested Butte and took her husband's name, Lohr) stayed on for ten years. Under her stewardship, more formal accounting practices were put into place and RMBL continued to thrive.

When Lohr decided to move on with her young family, Taber Allison was chosen to be next Director. Again, this was to be a one year appointment before another search was conducted. At that time, Dr. Ian Billick, a young scientist who had spent summers working in Gothic, was chosen for the slot. Billick took hold of the reigns and, with the backing of the Board of Trustees, propelled RMBL into the twenty-first century.

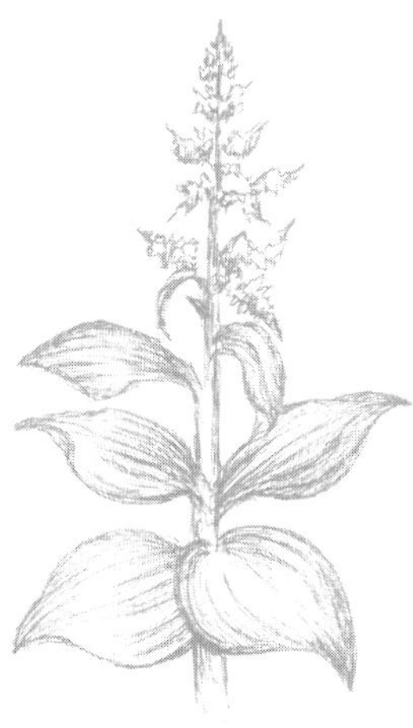

The Man Who Continues to Stay - billy barr

The place that is Gothic, Colorado and the organization that is the Rocky Mountain Biological Laboratory combine to be a powerful inducement to just BE there. Garwood Hall Judd, the Man Who Stayed, had come as a miner, evolved into a saloonkeeper and entrepreneur and remained as a colorful figure who regaled visitors with his tales. He, too, fell under the spell of the field station, choosing to help Dr. Johnson and the people he dubbed the "bugologists" who arrived with him each summer. Legend that he became, he has been joined in his attachment to both place and organization by another man, a contemporary legend – billy barr.[1]

It was summer of 1972 when billy came across a notice in the science department at Rutgers University for someone who had studied water chemistry to assist at the Rocky Mountain Biological Laboratory. He had no knowledge of the organization or the place it was located, but was looking for a summer job. He was, at the time, a junior at Rutgers and had spent the previous summer working on a dairy farm in Maine. Since making fifty cents an hour working on that farm was not an option for him again, he packed a bag and headed for the mountains of Colorado.

billy was a bit of a free spirit and, at the tender age of 21, at loose ends with his life. Without exactly knowing what he was looking for – or even that he was searching for anything – he found what he needed in Gothic, Colorado. Saying that he really liked the place, that it was somewhere he could live, he describes his arrival there as landing on solid ground after being afloat for many years.

An inner city boy scout in his youth, and having spent some time in Maine, billy did not find the transition to life in Gothic to be difficult. He passed the summer of 1972 working, doing water sampling, hiking and enjoying the mountains. When it came time to return to school in New Jersey, he could not bring himself to leave. He requested and received permission to take the semester off and settled into the area around the deserted mountain town. RMBL had denied his request to live in one of their cabins in exchange for services to the organization, so billy pitched a tent in which he lived until the snows came. Offered the free use of an old cabin by a man who owned mining rights to it, billy happily accepted and moved in until mid-December. By then he had decided he needed to finish his education, so he returned to New Jersey for the spring semester. Degree in hand in the spring of 1973, he worked in a factory just long enough to make money to return to Gothic. In late summer of that year, he headed west and made Gothic, Colorado his home.

For about eight years, billy lived in an old, deserted forestry shed. Approximately eight or ten foot square and barely constituting a shelter, it contained a small wood-burning stove to provide heat and a place to cook. Summers were spent working at RMBL, doing what needed doing. He would find employment in the kitchen or on construction or maintenance details, as well as on the fire-fighting crew. Although his degree was in Environmental Science, he never really worked "doing" science again. Winters were spent making trips to visit his family in New Jersey but mostly just "being" in his little patch of the Rocky Mountains. As there wasn't really a lot to do during the winter months, he began writing things down. The weather had always fascinated him. "Living out here, the weather plays such an important part of your life, even in summer. But in the winter it controls everything 'cause it's so intense and so overwhelming. So, I just wrote it down." The meteorological records kept by billy annually have become a major source of information for scientists studying the area.

Winters in Gothic were hard enough even for this intrepid mountain man to induce him to build himself a better, stronger, safer shelter. Having purchased a parcel of land adjacent to the Lab, he traded it for a different piece a little further up the road. On this land he built a house – a home – with the help of two friends. The house went up in the summer of 1980 and he spent winters finishing the inside. Well, almost – "Hell, it's not even finished yet!" he claims. Added to the basic house eventually was a room for watching movies and a greenhouse for growing fresh vegetables.

Although he has skied out for food and mail weekly, transporting delicate vegetables on cross country skies is not practical. They would freeze during the time it took to ski home. Electricity made night time movie watching possible and the internet came to stay, but billy would continue to cook and heat with a wood-burning stove.

At about the same time that billy built his house, RMBL found itself without a business manager. Since the bills were not getting paid and since billy was there, he offered to take over accounting for the organization, which at the time did not yet have a full-time director or year round office in Crested Butte. He claims that he was "good with numbers" and continued on as business manager for over thirty years.

When asked, billy admits that he played classical guitar for a while. He and two other Gothic residents performed chamber music pieces in summers, but with no one to play with in winters, that hobby went by the wayside. One that he picked up again after a several year hiatus is knitting. "I've been knitting for twenty five years, but I haven't knitted for twenty!" He knits mittens. They wear out a lot and need constant replacement in winters, so it is handy to know how to make them yourself.

To be spoken of with love and respect by as many people as is billy is quite an honor. Almost everyone at RMBL has a "billy story". One of his closest friends, Bobbi Peckarsky, cherishes letters he wrote for winters on end and has nothing but kind words for the man who has come to be like a brother to her. He'd impose welcome interruptions in her days of field work by stopping by with some tale that would make her laugh. One of their favorite things to do was to "publish" papers together. "There was barr and Peckarsky, and Peckarsky and barr. There were a whole series of them, all with a common theme: When it rains, the river goes up. When it doesn't rain, the river goes down. When it snows… well, you can take it from there… He often just made my day."

A quiet, self-effacing man, billy does not like to talk about himself. His friends and associates at RMBL tell tales of his years as a winter recluse and his exploits during the summers. He accepts the fact that people share stories about him, but is fully aware that facts are exaggerated and distorted through repeated telling – "the exaggerated eccentricities. I really get tired of that". True is the fact that he got caught in an avalanche one year and broke a few ribs; it could have been much worse. Also true is that he fell in the river one recent winter and bruised his back and kidney so badly he could hardly manage to leave his house – the same year that his movie projector died and his water system froze up, causing him to haul water from the office all winter. "That was tragic! All the most important things – my health, my water and my movies – were all gone that winter!" Yet he stayed. And he continues to stay.

"I don't think there's much accomplishment in doing what comes easy to you."

In his own words...

billy's Meteorological Observations, 2012[2]

"I've been keeping valid weather records for 38 years. I threw out the first year or two because I'd leave the thermometer out in the sun all day and it would get up to 71, or so. Not good. Over the last 38 winters, though, the average snowfall in the winter is 433 inches. Before last winter, it was around 438. Last winter we had only 253 inches. It was the third driest winter on record, but it was the earliest melt off date. So what you're combining [sic] is a dry winter, but the temperatures are so much warmer that the snow doesn't hold as long as in comparable winters.

"The deepest the snow has ever gotten in the Gothic area is 118 inches, which is just shy of ten feet. Last winter the deepest it got was 53 inches. Most winters it gets around two meters, which is 75, 80 inches, in that range like that."

A Cabin Called Home

"From 1972 through 1979, [a small cabin] is where I lived. It's not there anymore; it burned down in 93... It was about 8x10 or 10x12, I forget. The back room wasn't livable. The stove pipe was high off the roof so it wouldn't get covered by snow too much. The back room was unfit to even store wood, which was kept outside.

"I lived there for eight years. When I first got here, I'd lived in New Jersey all my life and I wasn't, let's say, overly stable. Living in that shack was wonderful. It was just great to be there. I didn't mind the fact that it was really small. Later in the winter... snow starts to build up; [first] it [gets] about half way up the door; the pipe on the roof was not quite standing up so much. What the snow does is insulate the cabin, because the cabin had no insulation. Later on ... most of the cabin is gone – the snow is so deep. The 77-78 winter was a heavy one. The winter before that is still the driest winter on record. Those were good winters to have a camera because you get the contrast. It was a really nice cabin!

"[I had to cut] dig-outs to dig down to the windows so I could get light in. That's just a tunnel to get light. There was a spring back there I'd have to dig out and get water from. That's how I got out – I would dig steps.

The cabin would be completely covered with snow, but it was so great to be here I didn't care. Now, I literally have dreams where I am still in the cabin. I wake up like panicked!

"I actually got creative with the steps. I made winding steps one year. The winters were long and I was bored."

Wildlife Encounters

Elk

"There was an elk one winter that got stuck up Copper Creek. It basically hung right along the river and would eat willows. This has happened a number of years and most years I'd go back later in the winter and there'd be just the bones of a dead animal. You can see its intestines looked like someone packed them with sawdust. There was nothing for them to eat. This one, however, actually made it; I saw it alive in the spring."

Foxes

"You should see the foxes... they will sit on the surface and all of a sudden start digging down through the snow, stand there and then pounce to get a pocket gopher, which they can sense from five feet above the ground. They just sit there like they are sleeping and their ears twitch back and forth; they are listening to things."

Birds

Rosy finches, pine grosbeaks and gray jays hang around Gothic in winters, along with mountain chickadees and Steller's jays. "The gray jays will come and ask you for food. The Steller's jays will wait for the gray jays to stash it in trees, and then they'll go get it."

Etcetera

Weasels - "They are excellent mousers – incredible, just wonderful!"

Voles - "They will get into your stuff, but they won't crap in it like mice do. So at least they are cleaner when they get into stuff".

Snowshoe rabbits – "In the winter, you never see them. They are really skittish and really difficult to run into. Normally, what you see is their feet in the air as they run away."

"I don't know how anything survives out here in the winters. It's just so hard to live."

Pine Martens and Skunks, Oh My….!!!

An old, rickety shack in the high country of Colorado is an invitation for refuge to anyone or anything in need of a little warmth and protection from the outside world - so discovered billy barr during his years of residency in his shack in Gothic. The pine marten that lived with him for three years – "That was really entertaining; really interesting." The skunk that didn't want to leave – "That was NOT either entertaining OR interesting!"

"In the shack where I lived there was a back room. I had a couple animals move into that. They could move into that back room and there was a post that held up a beam in the front room. The pine martin would come in the back room on the top shelf and run down that post. It would run around looking for food or mice or whatever, and then it would climb back up that pole and go back out. So it was not a big deal.

"The skunks are not big climbers. Their defense is that they don't smell really good sometimes. So this skunk would come in and be up on the shelf, and then it would go back and forth. Sometimes it would come in, go down the pole, and be stuck on the floor and wouldn't climb out. So I'd have to open the door and get a broom and get it to walk out the door and then I would whack it to give it the idea that if it comes in the house it gets whacked by a broom; so it shouldn't do that anymore. What happened once – 'cause they're not dumb, they are just fit to their niche – it came in, got down and I guess it realized it never got hit until it left - SO IT WOULDN'T BUDGE!. It sat there and sat there – I pushed it a little and it wouldn't move. So finally I pushed it with the broom and it flipped over. Well, that is how you kill a skunk. You get it on its back, like a porcupine. So that thing immediately jumped up and sprayed all over the house!

"In the winters, the cabin was completely covered in. There were no windows. I tried all the old tricks. I had a big can of tomato sauce, which I poured on the floor because people said tomato juice worked. That didn't work. I got kerosene and poured it on the floor. That didn't work. So I just had to leave the door open.

"That's the good part of the story. The bad part of the story is that they shouldn't be up and active in the winter. That thing wasn't in great shape. Some weeks later, I found it dead out front. But a skunk won't be active in the winter unless it's in trouble, unlike some animals.

"Now the one animal I did have live with me for three years that was wonderful was a pine marten. They actually aren't seen very often, but the fact is they're like the weasels; they're reasonably secretive because of predators. But they are very inquisitive so, if not threatened, they will come up to you, given time. This one moved into the back room. What it would do was go out and hunt every night. A couple times a night, I could hear it coming and going. In the daytime, it would mostly sleep in. It would come out sometimes and just sit around out front. It was a really pretty animal – most of it is fur, not flesh. It would get in above and then climb down the pole knocking everything off the shelf. Then it would climb back out. They are the size of a small cat.

"There was one day I was cutting wood outside and it was sitting next to me on a log. A vole ran by and the thing just watched it, then jumped off the log, hit it behind the neck, killed it, got back on the log and just sat there while the animal died. Then it jumped off the log, picked it up and walked off with it.

"The pine marten is a really good climber. It stays in the trees a lot, especially in summer, which is why people don't see them that much. It also hunts in the trees.

"This one was a really good neighbor. It was nice to have around. I did make some dumb mistakes, which you should never do. The first year, I fed it, which is how I got friendly with it, but still… One day I tried to experiment by seeing what it would do if I didn't give it food. It came up to me and I put my finger out. Yeah, dumb. It actually put its mouth over my finger and just barely put pressure on my finger, opened its mouth and took a step back. I swear it looked at me saying 'yeah, are you nuts?

"She lived with me for three years. It was a female. The third year she had a young one with her and that was awful. The young wouldn't leave and she didn't want it there; she would growl at it all night long – for months. It would wake me up constantly. Then they left and it was the last time she wintered there."

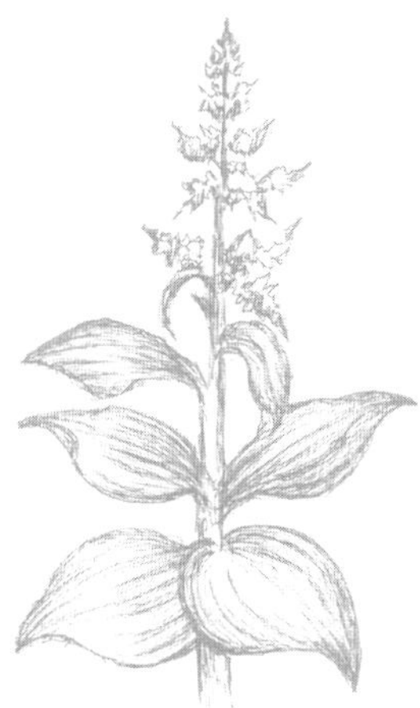

Tales Out of School... Bits and Pieces

No organization can run simply as a research center or educational complex or manufacturing facility. Behind the purpose of the place exists a logistical structure that is tasked with its own purpose – that of keeping operations running smoothly. These operations might include internal and external communications, feeding the hungry staff, cleaning buildings (and toilets), monitoring security, providing entertainment or even childcare. The list of support possibilities is almost endless.

In the usual course of things, the people who make up the support staff of an organization are skilled and paid. This is not necessarily true at the Rocky Mountain Biological Laboratory and has been historically untrue, for the most part.

The next few pages are devoted to highlighting some of the people and events that make RMBL a place to remember, that give it that sense of "community" that each and every person who spends time there describes. There are many more memories out there than there is room here to record, but it is hoped that these few will offer a glimpse inside the community that is RMBL.

The Road to RMBL

In his personal notes written over the years, Dr. John C. Johnson would record reminiscences of things gone by. This item was written in 1963 and reflects Johnson's continuing concern with the fragility of the environment surrounding Gothic.

I suppose I am the only person in the United States who ever went to the County Commissioners year after year, for 31 years to be exact, to keep a road poor that he had to travel nearly every day. The reason for the request was to preserve for generations, if possible, the beautiful and almost virgin territory around Gothic for professional and amateur biologists of the nation. Surely it is not selfish to hope that one area in Colorado can be largely set aside for biological study and research.

In spots, the road (1963) is narrow and dangerous, having about 180 curves. About one third of the distance is very muddy and slippery during and after a hard rain, scarcely passable even to the native skilled driver. Waiting for the road to dry out is a small price to pay to help keep the biotic communities around Gothic as natural as possible.

In spite of my requests the road was improved a bit each year. We are still doing our best to slow down the improvements in order to discourage the careless tourist who picks flowers and leaves picnic debris. Biologists and natural study enthusiasts of the future should have an unspoiled area to study the high mountain plants and animals of Colorado.[1]

Despite Dr. Johnson's repeated requests, the road continues to be improved regularly. Many more cars travel it annually than for which it was built. The drive is beautiful, but the wear and tear on the tiny hamlet of Gothic is evident to resident biologists and observant visitors. Small but precious animals are crushed under the wheels of heedless drivers; students and researchers alike flee from speed and dust; the air is polluted from the exhaust of so many vehicles. Bicyclists fear for their lungs and lives. Gothic has been designated a school zone, but many do not pay heed.

PLEASE DRIVE RESPECTFULLY
SLOW DOWN FOR SAFETY AND CONSIDERATION

Adventures in Dining

Back in the day… whatever day that may have been prior to electricity and refrigeration… keeping food fresh and safely edible was a challenge. This was true for people in general, but in a remote biological field station high in the Rocky Mountains it was particularly problematic – as was the simple act of cooking for a group. Westerners were accustomed to challenges; cowboys on cattle drives had chuck wagons to service their culinary needs; migratory miners packed their mules' sacks with canned and dried goods. The Rocky Mountain Biological Laboratory had its streams, both as a source of protein and refrigeration.

In 1929, the "Preliminary Announcement" of the second annual field season in Gothic advertised expenses for the six week program as $110.00. Of this, $50.00 was itemized for board – "excellent meals at $1.25 a day". An end of season accounting lists the actual cost of food items, purchased locally in Crested Butte, as $622.34, plus an additional $15.00 for coal and handling. Milk was an added expense, listed as $63.00, for a grand total of $700.34. The sixteen recorded paying participants would have contributed $800.00 for food. Considering that the Johnson family needed to eat, as did the other staff and researchers, this did not leave much room for creative dining. Basic it was then, and basic it would be for decades.

Handwritten menus in the archives of RMBL give some insight into the general fare offered at mealtime in the late 1930s and 1940s. Breakfasts were pretty standard – fruit, cereal, eggs, pancakes, toast and coffee. Offerings at lunch included such items as peanut butter sandwiches, an assortment of vegetable soups, sardines, fried bologna, scalloped oysters, macaroni and cheese and rice patties. Dinners would generally consist of a meatloaf or salmon loaf, stew or ham with potatoes, "a vegetable or two" and pie or pudding for dessert. One item that shows up regularly is canned fruit – especially peaches and apricots.

Bread was made in-house, as it was too costly and time consuming to drive to Crested Butte each day. The recipe for "**Gothic Bread**" has been retained for posterity.

1. Dissolve together: 1 ½ cups water @ 70-80 degrees, ½ tsp sugar, 3 tsp yeast (add 1 additional tsp yeast for sea level)
2. Heat on stove until hot to touch: 1 ½ cups milk, 3 T butter

Cool to room temperature
3. Mix together: 3 T sugar, 1 ½ T salt, 8 ½ cups flour
(reserve one for kneading)
4. Combine, stir, knead. Don't be afraid to beat. When the dough springs back, set in warm spot and let triple in bulk. Knead again, shape. Let rise until double. Cook at 350-375 degrees. (RMBL cooks often put the dough in a cloth and banged it against the counter with great force to mix it.)
5. To glaze bread: When almost cooked, coat with mixture of butter and sugar.

Finding someone willing to serve as a cook for summers was not easy, so the job often fell to the Johnson women. A fortuitous arrival in 1946 was preceded by the following telegram to Harriet Barclay: "MRS HIBLCOME FOR $25 EEK TRANSPORTATION AND MEALS ENROUT EXPECT DISSES WASHED AND TABLES WAITED ARRIVE GUNNISON TUESDAY 640 WIRE AT ONCE IF NOT OK (signed) MRS CHASE". [The errors are original.] Mildred Johnson recalled that John Johnson's 18 year-old niece, Edna Meyer, cooked for RMBL in 1934, and often served a soup she called "kitchen mystery".

Dorothy Johnson, who managed the dining hall on and off for many years, recalled that "it was not an easy job to be a cook up there [Gothic]. So many of the cooks who came ended up crying because the cake fell or because they put so many potatoes in the oven to bake and it takes a lot longer to bake umpteen potatoes, and here you are ready for supper and the potatoes aren't done and the cake fell. The cooks would cry!"

For seven years beginning in 1953, Chris Johnson was able to prevail upon a woman from Norman, Oklahoma to cook at RMBL. Louise Ward was in charge of freshmen dining at the University of Oklahoma and, according to Dorothy, she knew "how they were missing mom's food and what it was like to cook for kids away from home." Chris enticed her with tales of the beauty of the Colorado Rockies – but neglected to tell her that she would be cooking at very high altitude, on a coal stove, and that there was no refrigeration other than spring water in a tub. Louise rose to the challenge and returned to Gothic regularly for half a dozen years. Her husband would drive her out early in the summers and, while he was there, take it upon himself to repair any non-functioning washing machines. Her children would accompany her; in order to make life more comfortable for them, one year the RMBL students took up a collection to put a picture window in the cook's cabin. When she received a promotion at the University of Oklahoma, she was no longer able to return to Gothic.

Dorothy Johnson took control of the dining hall in 1969 and worked, unpaid, for years to make it run smoothly. In 1970, the cook who had been hired for the summer notified them that she had broken her leg and would, therefore, be unable to take up the job. Dorothy and Chris' daughter, Carol Johnson, had been prepared to be assistant cook that year, so she was promoted to head cook; her roommate, Jan Garton, who was there to serve as a helper in the kitchen, was made assistant cook. Dorothy spent the summer putting in twelve-hour days, teaching everyone in the kitchen how to cook.

Reminiscing about those years, Dorothy related, "We had these little recipes for six or ten people; we had to adapt them to serve sixty for breakfast and lunch and, then, for dinner mostly about eighty people. On Tuesdays and Thursdays, on seminar nights, we'd have anywhere from eighty to 125; we only had 126 plates! There would be days when we expected more people for supper than came – we'd ask each other if we remembered to send out the invitations for supper that night. Someone might reply, Well, I guess we forgot to stamp them!

"I remember one night, a summer when we did have a cook, she had told the kitchen staff to turn on the ovens at 3:00 and to put the salmon loaves in at 4:00. Well, the staff didn't get there until 4:00, so when it came time for supper we opened the ovens and the loaves were still really fluid. People started looking in the windows – they were hungry and it was time to serve. So, I said, we're changing the menu. Get out the skillets – we're serving creamed salmon on toast!"

There are tales of food orders being misconstrued. An order for fifty pounds of ground beef was understood by the provider to have an extra zero; five hundred pounds of hamburger meat was delivered to Gothic! Another time, someone mistook an order for radishes to mean fifty pounds. Fortunately, there were not that many radishes to be had in the area, so the problem solved itself.

Necessity being the proverbial mother of invention, a fair amount of creativity went into the cooking. For a while, homemade yogurt was prepared for breakfast. David and Bonnie Inouye recall baking all the bread for the dining hall the first summer they were there, using a bread mixer they constructed out of a large plastic trash can and an electric motor. Dorothy remembered the contraption as "a bucket with paddles" that would allow them to make three loaves of bread at a time. "We'd buy whole wheat flour for the bread. Later, I bought a gadget – a wonderful whirly thing – that would grind wheat. You could take the freshly ground whole wheat flour and use the paddle contraption and make three or four loaves at a time. On our way west [to Gothic for the summer], I'd stop and buy a hundred pounds of organic whole wheat flour. Once we got the gas ovens, it was a really big improvement."

Despite the challenges and the absence of pay, Dorothy loved the work and the connection it gave her with everyone in camp. "We had such fun in that kitchen!" she said.

There was no "official" dining hall for the first ten years or so, but in 1937 the construction of one was begun. It was completed in June of 1938 and served as dining room, lecture room and a recreation center. This building was replaced by the Enders dining hall in 1968, which served the community through 2014. Unfortunately this building had the distinct disadvantage of having been constructed in an avalanche zone so snow damage was not unusual. A new, modern dining and all-purpose facility replaced Enders Hall in 2015.

Until 1968, the lack of refrigeration was compensated for by the use of a large, concrete tub in the corner of the old dining room. A spring trickled down from above the Johnson House and was diverted into the tub. Five-gallon milk cans, as well as crocks and jars of food, were set in the tub with the water covering their bases. In this way, food items could be kept relatively cold.

The 1938 dining hall was cold and cramped; the Enders facility was larger. According to Dorothy Johnson, "You could tell it was designed by a man because there was a big kitchen. The sink, refrigerator and stove were far apart and it took some manual lifting to move the dirty dishes from the return window to the dishwasher." But, it was a great improvement because there was gas to cook on and nice, big ovens. There was also a root cellar, but it was rarely used.

During the 1970s, RMBL at last hired a full time Director who worked diligently with the Board of Directors to help assure that the organization operated with more of a financial cushion. One of the things that was done was to raise fees so that the quality of food could be improved. Professional cooks – meaning those who cook for a living - have been *de rigeur* since the turn of the twenty-first century. One such chef, Chad Carter, worked in the RMBL kitchen for five years, from 2002 through 2007, and is remembered fondly for the quality of his duck dishes. Chad enjoyed his time in Gothic so much that he went on to graduate school for food science. Along the way, he became interested in and studied "Protein, Amino Acid, Proximate Composition and Baking Characteristics of Maya Nut Flour". Once a RMBL chef, also a RMBL researcher!

The new community center, completed in 2014 and opened in 2015, serves as RMBL's dining hall and seminar room, as well as an inviting space for major programs. The facility is a testament to the growing reputation of the Rocky Mountain Biological Laboratory, funded by a capital campaign which was supported by many who simply want to see RMBL thrive. The new building will serve the needs of the Gothic community for generations to come.

Feral Kids - Growing up in Gothic

One of the most beloved aspects of the Rocky Mountain Biological Laboratory has always been the inclusion of families. Despite the lack of modern housing and facilities, the ability to live in a cabin, cook on a coal or wood stove and have outhouse and shower facilities relatively nearby appealed to many scientists who wanted or needed to spend their summers doing field research but did not want to be separated from their wives or husbands and children. Also, the proximity of other families helped to assure that children were safe and watched after, if not entertained. The one descriptive word that is repeated by anyone who has lived in Gothic, if only for a short while, is "community". And part of the reason for the sense of community is and has always been the presence of families.

The first children to summer in Gothic, and for whom it has always been a second home, were the Johnsons. Clea, the youngest, has always remembered her summer days there – a bit of work, a lot of play. What could be better than the mountains of Colorado for unsupervised summer camp? There were streams to ford, frogs and snakes to catch and examine, butterflies to chase, horses to ride, trees to climb. The ways to fill one's days were endless.

The Remington family began their summers together in Gothic in 1954 and, taking advantage of RMBL's policy that allowed investigators to build cabins and hold a lease on them for up to forty years, they arranged for the construction of their own cabin in 1956. It was, and still is, called the Remington cabin. Janna Remington and her brothers were quite young and recall life as a fairy tale. There were not many rules, other than to stay away from the study areas. Chores consisted of things like daily hikes to the spring to fill up the water jugs. Even those were adventures, shared with the children of other scientists.

"Picture us kids," Janna recalls. "We were gone all day. We'd come home for lunch, but we were pretty much out fishing, horseback riding, building forts up in the woods. We played and fished around the East River. My brother Eric was famous for his fishing – he'd catch his limit of twelve fish in less than an hour. For the most part, the kids were allowed to run free. Nobody ever worried about us drowning or falling off a mountain or anything."

Horses were an important part of many summers. It was fairly inexpensive to lease a horse for the summer, so some children experienced the freedom of life on horseback. The horses were let loose at night in a large pasture, but each morning the lucky ones who were allowed horses would head for the pasture in search of their mounts. Once captured, the creatures were used as legs for the happy kids astride their backs.

There was room for creativity, too. According to Janna, performing plays was a special part of each summer. "I and Anne Baker and Bobbie Novitski would write the plays. Then, we would get all our brothers and sisters to act in them. We'd invite the parents – I think we may have even charged them a nickel or something to attend. We made costumes; we learned lines and everything. We built a little stage. It was great fun."

Andy and Christine Beattie have fond memories of the times their children spent in Gothic in the 1970s. "One of the great things about having children here is that there is such a great amount of freedom. It may have changed a bit, but normally it was possible for children to grow up by themselves and do what they wanted to do." Their son, who was eleven the last year he spent in Gothic, was particularly fond of snakes. In Gothic, he could pick them up and not worry about being bitten because the ones that inhabit the area are not venomous. In Australia, where they moved to live, this is not the case.

Bobbi Peckarsky and her husband brought both of their children to Gothic as babies – and as teenagers. "This is a cool place to be with a baby. It is even a better place to be with a teenager", she claims. "They learn how to be without stuff. My kids were born in the 80s, so there weren't cell phones; but there were Game Boys. They almost never used them. They came out here and built forts and made little hiding places in areas and threw stones in the rivers." Her children were infants in 1984 and 1986 when they lived in the Barclay Cabin. There was no hot water, but there was an "awesome six or eight burner wood-burning cook stove, with a bread warming oven. It was the source of heat and we cooked on it." They would also heat water for baths. "We had a big metal tub which we would put on the floor in the living room and pour hot water into it that had been heated on our stove. Our two little kids would sit in the rub-a-dub-dub with their rubber duckies! We also built them a sandbox and all their friends came to play in it. They got to the point where they didn't want to leave their friends."

Friendships formed during summers in Gothic were annually renewed and became, for many, life-long connections. This was as true for the children as it was for their parents. One would think that so many months spent basking in surroundings infused with research might have infected the kids with a love of science. Love it they might, but choose it as careers most did not. Many of their lives moved into vastly different areas, ranging from law to engineering to ballet and beyond. But somewhere in the hearts and minds of those who enjoyed their childhood summers there, Gothic is "home".

Gothic to Crested Butte Run, Walk or Crawl One-Third Marathon

One summer evening in 1969, the editor of the local Crested Butte newspaper, *The Chronicle*, decided to pay a visit to Gothic. He did not call ahead, as there were no phones, nor think about making an appointment in advance. He simply headed up the mountain after getting the paper out, arriving quite late, after dark.

George Sibley did not realize that if one were going to walk around Gothic after dark, a flashlight was a necessity. "I was out on the road, stumbling around, and somebody came along with a flashlight. I asked where I could find Chris Johnson and the person was kind enough to take me to him. It was a typical evening for Chris and Dorothy. There were probably twenty-five people in their little McLeod Cabin. I walked out of the darkness into bright lights and all these people. It was kind of stunning." According to Sibley, he and Chris Johnson had previously met. Chris' son, Doug, remembers the story slightly differently.

"On a typical Friday and Saturday evening there would be gettogethers in the McLeod Cabin. Mom would have something to eat, coffee and tea; people would sit around and talk. This one evening, there was a knock at the door. Someone opened it and George walked in. He is a really friendly guy, greeted everyone and was invited in. Mom offered him a cup of coffee and asked him to sit down. We were all talking for a while and pretty soon people started asking each other if they knew who he was. We realized after a while that nobody knew this person! Introductions were finally made all around."

That evening remains in the memories of many people who were present partly because of something that came out of it. Chris Johnson thought that it would be a good idea if RMBL and the town did something jointly. Marching in the Fourth of July parade on Elk Avenue in Crested Butte, clothed only in skunk cabbage leaves, began as an annual tradition for RMBL students at about that time, but Chris and George decided to work on something else. They agreed that a "marathon" between the two towns would be a great idea. The fact that the distance is less than marathon length did not deter them from calling it that at first. It would later become known as the "Gothic to Crested Butte Run, Walk or Crawl One-Third Marathon" and has been run every Fourth of July morning since 1969. Gothic to Crested Butte Run, Walk or Crawl One-Third Marathon #

As Dorothy Johnson remembered it, the prizes offered that first year were simply for crossing the finish line. If you were of legal drinking age, you received a beer. If you were underage, a soda pop was the reward. And if you could prove you were the oldest person to run, your prize was a one year subscription to the *Crested Butte Chronicle*.That first year, the winner of the subscription was Abby Enders, wife of Director Bob Enders.

Many people who have a long history with RMBL also recall particular experiences they had with the "marathon".

Ken Armitage, who first came to work at RMBL in 1962, was an enthusiastic supporter of the race. He first ran in it when he was fifty years old, and continued for another fifteen years, or so. "One year, as I was mrunning, I was talking to some guy who was from Topeka, Kansas. I am from Lawrence, Kansas and we were jogging along together. I asked him when he got here; yesterday, he said! I felt like I wanted to push him over the hill, as I was laboring along!"

The official name of the marathon is the 'Run, Walk or Crawl One-Third Marathon'. Armitage claims that, "as far as I know I'm the only one who, on the same Fourth of July, ran, walked and crawled! We got down near the finish line; I stopped, got down on the ground and crawled across the line, just for fun. Another year, George Sibley and I ran up Elk Avenue with banners saying, 'Marmots Run for Fun'."

Not everyone showed their support for the race by actually running in it. Dorothy Johnson and other women would frequently walk the distance, escorting children along the way. Former part-time Director Andy Beattie enjoyed dressing up in costume to start the marathon off in the mornings. "I said, ready, get set, go! – and set them off. Then, we went back up into Treasury Cabin and broke open a bottle of champagne. You wouldn't get me running down there!"

In 2011, Dorothy Johnson recorded in a letter that, for many years, she and Katie Armitage, along with "some other Gothic residents", had kept the marathon going so that their sons could run. George Sibley had sold his newspaper, but they worked with other sponsors in Crested Butte, including the local bookstore, Heg's Place, and the race became increasingly more popular. "One year, to our disbelief, RMBL committee member Dr. Stan Dodson announced, 'People will pay to enter this race!' And they did! No longer just a fun run – but still fun – hundreds of entrants paid to run, and the 1/3 Marathon continues to be a major fundraiser for RMBL's library and other worthy projects."

Whether as participant, sponsor, behind-the-scenes coordinator or observer, many people continue to enjoy the annual event that began so humbly in the minds of two new friends in 1969.

Birthing a Babe - The Old Fashioned Way

One of the most frequently recounted stories of life in Gothic is that of George and Barbara Sibley delivering their own child in the middle of winter, with no one around to help them. As the common story goes, Barbara went into labor in their cabin when they were serving as winter caretakers, George delivered the child and they kept it safe and healthy until spring, when they were able to emerge from the winter's isolation.

Lovely tale. It is, however, not true – except for the part about their child being born in Gothic. The story unfolds…

The year was 1974, but the month was August. The Sibleys were living in a cabin in Gothic, working around the camp. Barbara was heavily pregnant with their daughter. The plan was for them to drive to Gunnison when Barbara went into labor or, alternatively, for their physician to drive to Gothic to deliver the child. There was some justified concern about Highway 135, the road to Gunnison, being impassable due to construction, so they asked the doctor for some basic child birthing instruction. Telling them that one cannot learn to deliver a child in a few hours time, the doctor suggested they go to Gunnison and have labor induced. Unwilling to do so, they took their chances. As these things go, when Barbara's labor began it was during the night; the road was too challenging for them to drive; and the doctor was on duty at the hospital, so unable to make the trip.

During those years, there was often a volunteer doctor in residence in Gothic during the summer term. The physician this particular year was, according to Dorothy Johnson, Dr. Hull, the son of the original doctor for whom the doctor's cabin is named. When Barbara went into labor and it became apparent that there would be no trip to town, George contacted Dorothy, who contacted Dr. Hull. He arrived at the cabin just as little Sarah was being born and took over the rest of the delivery.

In the words of George Sibley, "Anyone who has ever been present at a birthing knows that what the guy does is sort of cheerleading. We are cheerleaders for the victory of the team. Barbara delivered the baby and I assisted."

In the morning, Dorothy recalls receiving a radio communication from George saying, "Sarah Avery Sibley says 'Good morning world!'"

So, a healthy child was born in Gothic, with the assistance of the local community. The Sibleys did spend the winter of 1974-75 in Gothic as caretakers, but had quite a scare when little Sarah contracted bacterial meningitis on a trip to town. A friend with a snowcat came to their rescue and they were able to obtain timely medical care. But that incident, and the challenge of living in a cabin measuring 16x20 feet with two small children, convinced them that their years as winter caretakers were over.

Winter's Tales - All's Well that Ends Well

(With apologies to the Bard)

Winter comes early to Gothic, Colorado, each year a little differently. October might find the little town snowed in, or it might be possible in November to drive through it up to higher country. It is not unusual for unenlightened individuals, strangers to the area, to satisfy their curiosity and attempt passage at a time when it might not be a good idea. Hikers might head over from Aspen, miscalculate the time it takes and be overcome by darkness or a sudden snow squall. Unanticipated ice might keep a vehicle from completing a journey. And, of course, accidents do happen.

RMBL winter caretakers, billy barr and anyone who has spent some winter time in and around Gothic have tales to tell of encounters with the lost.

November 1973.

billy barr had settled into his second winter in Gothic, living in a tiny, uninsulated old shack. George and Barbara Sibley, the first Rocky Mountain Biological winter caretakers, were at home in their cabin for their third winter in Gothic. Kevin Donovan, local builder and sometime RMBL contractor, was living in a tipi with his girlfriend, Shorti, just above Gothic. It was a Tuesday night – that much is clear because billy always baked bread on Tuesdays and invited his friends to share it. It wasn't exactly bread as one might know it today. His wood-fueled oven wouldn't heat to more than about 250 degrees Fahrenheit, so the resulting mass was somewhat doughy – "a cross between bread and beer". But it was a reason for coming together.

Memories diverge a bit – it may have been snowing or possibly not, but there was definitely snow on the ground this particular Tuesday evening. It was, however, dark. Everyone was startled by a sudden rap on the window. The individual intruding on the little gathering was a young man dressed in "bankers' pants and loafers". Driving a Honda Civic, he had attempted to travel from Crested Butte to Carbondale, ostensibly to visit relatives. He had consulted a map and determined that the route through Gothic and Schofield Park past the Devil's Punchbowl and Marble was the most direct. Obviously, it was unfamiliar territory to him. He managed to make it to the Crystal River, but while trying to ford the river, the car hit a rock which tore out the oil pan. So much for the Civic. This could easily have been laughed about and chalked up to costly foolishness if it weren't for the fact that the man was traveling with his young daughter.

The little girl was somewhat older than one, but younger than two years old – way too young to be able to walk with her father to safety. Since he was wearing city shoes, it was equally impossible for the man to carry her and maintain secure footing on the icy, snowy road. What to do? It was late afternoon and the child was not yet tired, so the man played with the child until she finally fell asleep. He then wrapped her in blankets and secured her about forty feet back in an old mining shaft that he discovered. Then he walked to Gothic. Alone. In the dark.

Finding billy's cabin occupied, he rapped on the window and asked for help. After going to Sibley's cabin to radio for help, the four men piled into Sibley's Volkswagen squareback, which had chains on the tires. They were able to traverse Schofield Pass and had no difficulty driving through Schofield Park to the river and the mine shaft. Once there, the four of them crowded into the old mine where they found the pile of blankets - and the sweet, sleeping little girl.

Returning to Gothic with the child, Sibley and billy then drove the two stranded wayfarers to Crested Butte. "George and billy enjoyed a beer while the man called a relative in Gunnison who owned a towing business". The relative took charge of the man and child that evening and retrieved his car for him next day.

A 1970s visit to billy barr

Nick Waser first came to Gothic the same year as billy barr and the two men became lifelong friends. During the winter of 1974 or 1975, Nick and Mary Price decided to make a trip to Gothic and visit billy. They skied out to Gothic on a day when the temperature hovered around fifteen degrees below zero, Fahrenheit, recalling that they had never felt so cold in all their lives! The one room shack that billy was living in had no insulation, one tiny window, a lean-to room for coal – and a pine marten in residence. When they arrived, billy was not there. He had skied over to Conundrum hot springs for the day. Having made the effort to get there, Nick and Mary waited for him in the freezing cold. billy returned at last and rewarded their patience by baking a pie – a wonderful fruit pie made with peaches or apricots he had stored!

Sometime in the late 1980s

Susan Lohr was Director of the Rocky Mountain Biological Laboratory. She had been hired in 1986 and spent two winters living in Crested Butte before moving full time, year-round to Gothic. She was not totally alone there in winters, since billy barr lived up the road, and winter caretakers occupied a cabin. During September and October, however, billy would often be gone visiting family and the caretakers would not yet have arrived. So she would be, indeed, alone.

One evening during one of these alone times, Susan was sitting in front of the big window in Gates cabin, where she lived, "knitting or quilting or something". Darkness had already fallen and, all of a sudden, a man's face appeared in the window, illuminated from below by a flashlight. She could clearly see a gun in the man's hand. She was quite rightly terrified! However, Susan had spent many years traveling in South America and, as a rule, could conquer her fear. She pulled herself together and went to the door. Calling to the man, she asked who was there. The man approached the door and Susan told him to put his gun away. He did so and she saw that he was accompanied by a young woman, dressed in shorts and a tee shirt. The girl was blue and shivering from the cold.

The unimaginative ending of this potential horror story is that, earlier in the day, the couple had driven to Emerald Lake above Gothic to view the autumn leaves. Their car had broken down or become stuck, so they decided to walk out. By the time they reached Gothic, they were afraid they would die from exposure. They meant Susan no harm, but had seen the light in her cabin and headed that way hoping to find assistance. Wielding the gun was theatrical, but they were frightened. Susan still had her Subaru parked in Gothic, as the winter snows had not yet closed the road, so she drove them to Crested Butte and returned home – safely and alone.

The Early 1990s

Alison Brody, who had come to Gothic for the first time in 1981 to work with Ken Armitage on his marmot program, applied to be a winter caretaker with her husband during the winters of 1990-91 and 1991-92. She was writing her dissertation and he was writing a book. Several other people were wintering there, including the Director, Susan Lohr, billy barr and another couple for a few months, but for the most part it was pretty isolated. One morning, she and her husband awoke to the smell of smoke. The old shack that had housed billy barr for many years had burned to the ground during the night. Fortunately, no one was living there as billy had built and moved into his new home just outside RMBL's perimeter, across the road from the old shack. As it turned out, a man had skied out and spent the night in the shack. Apparently, he had had an argument with his girlfriend, drank too much and ended up in Gothic. It was not known if he had intentionally set fire to the building, or if he had tried to build a fire, or had tossed a cigarette butt onto the dry wood, but he did not stick around to explain.

Alison needed to go to Crested Butte anyway that morning, so she skied out. Along the way, she noticed boot tracks in the snow, heading off into the woods. She notified the police of both the fire and the tracks; the police were able to locate the man and arrested him.

The frightening part of this tale is not so much that it happened, but that no one knew anything until morning. There were only three residents in Gothic that night – Alison, her husband and billy. Even her dog failed to alert them to anything. The man had been able to arrive during the night and leave in the morning without anyone being aware of his presence, leaving behind the ashes of a building.

Due to the high altitude location of Gothic and Mother Nature's sense of humor, it is not impossible for there to be "winter" in the summer months….

Summer 1947

Mildred Johnson enjoyed recounting the time when twenty-three-year-old Chris Johnson arrived in Gothic with three college friends, strapping young men with big appetites. "We had a lot of work to do," said Mildred, "so I'd made a big pot of stew - to last the full week." She was rather surprised when Chris and his friends quickly made short work of it. So, they made a special trip to town, "stocking up on canned peas, corn, tomatoes and peaches, and extra bread and milk." It then proceeded to snow. The road was impassible and another trip to town was out of the question. It was uncomfortably cold in the Johnson House because the chinks between the logs had fallen out, and the freezing air blew in. The food ran perilously low. Resourcefulness will out, however. A search of the larder revealed "tins of rice and dried lima beans that had been stored from the previous year." The food was, thankfully, edible and no one starved.

June 1969

A blanket of fresh snow surprised Gothic on June 25, 1969, amounting to nearly a foot of the white stuff and putting a halt to planned research and education projects. The roads were treacherous and no one attempted automobile travel. Weather forecasters predicted the snow would linger for several days, along with unusually low temperatures. How to entertain the "troops"?

Ever creative, the kitchen crew brought a tiny evergreen tree into the dining hall. They worked to decorate it with upside cuttings from cardboard egg cartons, resembling bells and hung with string, as well as orange bows made from the tape used to mark boundaries of study sites. While people were engaged in this little project, into the dining hall walked several strangers. They were cold, wet to the bone and rather desperate. Participants in a well-known company's outdoor survival program, they had been caught with minimal gear and supplies in the area of Schofield Pass when the storm hit. Insufficiently prepared for the turn in the weather, the men had started walking toward Crested Butte. Their progress was slow, breaking trail in the deep snow, and they were fearful they would not make it to town. Coming over the hill into Gothic, they were greatly relieved to see lights burning. They had no idea there was any human habitation in the area – their map had identified Gothic as a ghost town!

The RMBL community welcomed them, provided them with dry clothing, invited them to join in dining hall meals and supplied wool blankets and bunks for several nights' rest. In return, the men spent three days helping to chop wood before they left with heartfelt thanks and an appreciation for the vagaries of mountain weather – as well as for the Gothic community.

Dorothy Johnson recalled, "We all realized the men would never have survived the next 8.3 miles to Crested Butte if Gothic had still been a ghost town on that snowy June night of 1969."

Acquiring the "Extras"

For decades, the Rocky Mountain Biological Laboratory operated with nary a modern convenience. Toilets were outhouses. Showers ran cold water and were located outside cabins. Laundry was sent to Crested Butte, for lack of facilities in Gothic. Light was provided by kerosene lanterns. Cabins were heated by wood or coal stoves. Communication facilities were non-existent, as was television. Even AM-FM radio reception was problematic. Residing there was akin to camping out in the wilderness, albeit in cabins.

By the end of the twentieth century, "modernization" was in full swing. Hot and cold running water was made available in many cabins, as well as in the community shower facilities. Composting toilets and upgraded outhouses with solid concrete vaults were erected throughout the townsite. The need to conserve precious water resources (among other things) continues to inhibit the installation of flush toilets, even in the newest buildings. Electricity has been installed in every cabin - and works! Land line telephones are commonplace. Most cabins are better insulated, and wood burning or coal stoves are becoming rare. The twenty-first century has brought some additional changes to Gothic, such as internet service and sporadic cell phone signal availability. Living in Gothic is much more comfortable than it was thirty years ago. The story of how things got from there to here is, as one might come to expect, one of ingenuity and cooperation.

A telephone line had connected Crested Butte to Gunnison in 1882, but for more than a hundred years it went no further.[1] In the 1980s, some of the people working in Gothic during the summers began to express interest in having telephones there. Prior to that time, contact with the outside world required a trip to payphones at the ski area or to Tony's Conoco in Crested Butte, where RMBL people could make phone calls and collect packages. Inquiries made to the telephone company returned information that was not encouraging – estimates for the cost of running and burying phone lines to Gothic were for far more than the research station could afford. Dr. David Inouye decided to take up the challenge.

Appling for and receiving a "little bit of money" from the Boettcher Foundation in Denver, Colorado, to purchase the cable, David sought help to accomplish the work from the people summering in Gothic. There were, at the time, volunteers from the Earthwatch Institute working at RMBL.[2] One of the volunteers was an ex-Marine by the name of "Moose", who had worked with a telephone company. He liked it in Gothic and told David that, if his expenses were paid, he would return the following summer and help run the phone line.

Next summer, true to his word, Moose returned. He and David took it upon themselves to get the job done, with help from billy barr and perhaps a few stragglers. RMBL received permission from the electric association to utilize their poles for the telephone cable. The distance between the poles was long enough that a strong cable was required to support the length of the run. Using a combination of David's jeep and motorcycles, they ran the cable underneath the power lines. At one point, they "hired a guy with a mule to help drag the phone line across the river". Moose and David climbed every one of the power poles from their start to their destination, "to put bolts on to hold the brackets that would hold up the cable". Then, they stretched the cable.

Once the phone line was actually in Gothic, Moose showed David how to do the internal wiring between the buildings. Those phone lines were relied on by RMBL and those scientists who chose to have a phone in their cabins for about ten years, gradually diminishing in number from the original twelve as they deteriorated or were otherwise damaged.

In the earliest days, none of the cabins had running water, so pails of water had to be hauled from the spring near the Johnson House, or from pipes that brought water down from Gothic Mountain and across the East River. Gradually, running cold water was installed in several of the cabins in the lower part of the townsite. In the 1960s, another scientist-initiated project brought water to cabins that had been built on the eastern hill in Gothic. Paul and Anne Ehrlich, as well as the Remingtons, the Novitskis, the Bakers and the Wagners had built their cabins on the hill leading up from the base area of the RMBL buildings. There was no plan for water to be extended up the hill to these cabins so the group got together and decided to underwrite the cost of having it done. Unlike David Inouye, who physically took on the job of running phone lines, they decided to hire a contractor. Had they not pursued this, it is uncertain how long they would have continued hauling water with buckets and milk cans.

Construction projects of all kinds were a part of summer life in Gothic. Not all scientists had the time or the interest or the ability to contribute of their labor; many did, and others contributed of their funds. Summer students and staff were deeply involved in building and maintaining the physical plant. With the advent of building and other codes it became necessary for much of the work to be contracted out. But for those who have had a role in building the facility, it is a matter of pride to have been involved.

Building and Re-Building

Kevin Donovan Joe Savoy

It was 1972, the year that many students who would become regulars first arrived in Gothic. Kevin Donovan was among that group, but he was not destined to be a research scientist. His future lay in construction.

As an undergraduate at Haverford College in Pennsylvania, Kevin decided one summer that he wanted to attend a biology field program in Wyoming. His advisor, a long-time friend and colleague of former RMBL Director Dr. Robert Enders at Swarthmore College, suggested instead that he go to the Rocky Mountain Biological Laboratory, which did not appeal to him at first. Looking at the pen-and-ink drawing of it, he thought that it looked "cold and rocky". But in the end he agreed and went "kicking and screaming" to Gothic, Colorado.

The reality of RMBL turned out to be much better than the anticipation of it. Kevin shared a cabin with billy barr – the Old Town Hall – and was pleasantly surprised by the vegetation in the area and the welcome he received from those he came across. After graduating from Haverford the following spring, he decided to take a short hiatus from life and returned to Gothic in September, setting up his hand-sewn teepee as his home. Attracted to the area in general, Kevin worked as a research assistant to Dr. David Inouye in 1974 and "did a bunch of field botany stuff over the next couple years". But he also began to work in construction. He opened a business in Virginia and returned every several years to build something in Gothic. The Gates Cabin, the Rogers-Boggs Cabin and the new Johnson Lab were his creations, as were the cabins named New Savoy, Calder, Teocalli and Maroon. Additionally, he made repairs and built an addition to the Levi cabin after an avalanche. His first full project in the area was billy barr's cabin, which he built along with billy and some friends in 1980.

RMBL had, from the beginning, been rather in a bind concerning its buildings, some of which dated from the late nineteenth century mining days. Their condition only continued to worsen with time. As many of the older buildings fell into varying states of disrepair, there was never a lot of money to build new ones. Also, there was a certain amount of pride in the community regarding the "original" mining town atmosphere of Gothic.

New and modern just wasn't in the cards. Historic preservation and renovation were going to be the answer.

When the Johnsons and their colleagues were settling into Gothic in the 1920s and 1930s, they made do with the rickety old buildings that Garwood Judd had helped them to acquire. As those buildings began to come apart, the wood from them was set aside to be recycled into new cabins, along with logs skimmed from the mountain sides. Joe Savoy and his wife, Mary, had emigrated from their native Italy, arriving in Crested Butte in 1920. Joe became a Crested Butte resident of legendary strength and ingenuity who worked in the mine in winters and for RMBL during summers, where some referred to him as "the Paul Bunyan of Crested Butte". Between 1929 and 1941, when he died in a mining accident, he was responsible for much (if not all) of the new construction in Gothic, including the cabin Dr. John C. Johnson had built for himself and his young family. Using only manpower and horsepower, Savoy harvested fallen trees for his building projects, hauling them down primarily from Snodgrass Mountain. He then carefully fit the logs together using only hand tools and clever methods. In the later decades of the twentieth century, several of the cabins he built would be deemed worthy of preserving, along with some from the later mining era. Kevin Donovan was the builder who was asked to make most of it happen.

When queried as to how he became interested in historic preservation, Kevin's response is, "Being asked helped a lot. It wasn't a weird fit or anything. It is a different kind of building; it is definitely rustic. The structure is held up differently. Some of the techniques are just not used on modern construction – like chinking and fitting the logs together." Many of the challenges to reconstruction come from modern building codes. There were no such restrictions prior to circa the 1980s in Gunnison County, so nearly all of the buildings were erected pre-code. Staircases, in particular, proved to be problematic for reconstruction. "Typically, the old staircases don't meet any aspect of a code staircase, so we pretty much have to take them out and start from scratch. All of the places we've redone have had a second floor, except for the Swallow's Nest".

Almost everything he has worked on has required a new foundation. The outside appearance is one of authenticity, but under the rocks there is a full concrete foundation. The Johnson House, which is now a dormitory but was built for the founder and his family, at one time had a large stone fireplace. It had to be removed because "it was dragging the whole house down the hill. The house, forty feet long, bent twelve inches out of straight in the middle. That's pretty impressive!" Originally built on a "bundle of posts", it continues to look that way but there are thousands of pounds of concrete in the ground that are not visible. The one-time famous catwalk that connected the two sides of the second floor was nixed by the county, much to the disappointment of those who knew the house in its younger days.

In 1995, Kevin moved himself and his business to Crested Butte, partly to be close to the woman who would become his wife. Sonda was working in the kitchen at Gothic when they first met, but she wore many hats during her ten years with RMBL, including that of Director.

The cabin built by Harriet and B.D. Barclay is a bit legendary because of its dimensions. Both of them were quite short and the ceilings were built notoriously low, as were the window sills. The Barclay cabin was among the historic reconstructions done by Donovan. Word about town is that he raised the ceiling, but that is not true. Some thought was put into lowering the floors to create more headroom, but that was not done either. "Finally, we just kept them the way they were. The lowest part is about six feet two inches – not really a head banger but a lot lower than most ceilings."

The surviving Savoy-built structures preserved by Donovan and others include the Johnson House, Barclay Cabin and the Vera Adams Johnson Laboratory. Two mining properties that were built around 1914, during a false mining resurgence, have also been rebuilt by him – Mammal Lab and Ore House. Donovan did the preservation work on the Richards Cabin, which was built in the early 1930s at the behest of RMBL scientists, probably with the assistance of Joe Savoy. Kevin also was the primary contractor on the preservation of the McLeod Cabin (possibly a Savoy project) and the Town Hall, and he consulted on the preservation of the Swallow's Nest.

The new Visitors Center which opened in 2013 is built with logs salvaged from the first laboratory built by RMBL, which was constructed in the mid-1930s by Joe Savoy and named in memory of Vera Adams Johnson. The structure is essentially the frame built by Savoy; it's an impressive visual lesson in fitting massive logs together, all work done without motorized machinery. Kevin Donovan and his crew were responsible for establishing the recreated building on its new, concrete foundation. Every care has been taken by them to ensure that all completed renovations are outwardly, at least, authentic to the original buildings – even down to preserving initials and names etched into original chinking.

Kevin is very complimentary of Joe Savoy and the work done by him. "I would love to have met the person. Looking at his buildings and working on them, I've sort of felt that he's always been kind of there. It must be fun for him to have done all this stuff and be able to look at what he did. I would like him to know that somebody cared about his buildings. His original workmanship is pretty good. It's mountain functional – they skidded logs off the top of Snodgrass; they used what they had and picked up local rocks." An understandable extension of Donovan's respect for Savoy is a bit of mental competition. "Over the years, I've said to him in my mind - well, Joe, I've now built more buildings here than you have. He had the record for building all the structures, but now it's kind of fun because I've worked on all his buildings, plus a bunch of others that I've designed or built."

All the News... And Then Some

A prolific writer, Dr. John C. Johnson began a tradition at the Rocky Mountain Biological Laboratory of near constant communication with his fellow founders, as well as with incoming and continuing faculty. A stream of letters, many of which still rest within the Johnson family papers, were exchanged between Johnson and almost anyone associated with the organization. He penned notes to students, welcoming them and explaining such mundane things as what kind of luggage to bring and the proper way to dress in the field. He wrote newspaper articles and kept a detailed personal journal. He authored letters to the RMBL membership, and seemed to really enjoy the process of putting pen to paper. Thankfully for the historian, or any interested party, the tradition of putting things in writing, and the plethora of such material, did not end with him.

The Gothic News, a summer "in-house newsletter", was created and distributed for decades by interested individuals at Gothic.[1] The brain child of Katie Armitage, it began in the mid-1970s as an activity for children. Eventually, it evolved into a somewhat regular paper, issued about every ten days throughout the summers. Katie Armitage worked to include calendar listings and summaries of lectures, which she claims she "wrote up with mixed accuracy and results". After a few years, Dorothy took responsibility as editor and, in Katie's words, "*The Gothic News* became a real record of activities".

Dorothy was quite proud of the impact she felt *The Gothic News* had, given that the newsletter existed before telephones were common at RMBL and before email or the internet became effective communication tools. It is also clear from Dorothy's stories, recounted to family members and others, that she thoroughly enjoyed the process of putting the paper together and distributing it within the Gothic community, as well as to RMBL friends in the Gunnison valley. Unpaid volunteers not only worked to create the newsletter, but they "even hand-delivered it to neighbors on nearby mountainsides down valley and to interested friends in Crested Butte and Gunnison. Chris [Johnson] put it into the computer and wrote about Gothic's history. Four of us collected news, wrote and/or edited. Artist Lorene Calder's illustrations graced the top of page one, which also listed the schedule of coming events; Jean Cairns was a reporter gathering news everywhere; Katie Armitage or I [Dorothy] or a fill-in volunteer wrote summaries of the Tuesday and Thursday night seminars. There was a column for nature notes about unusual sightings of rare animals, birds, etc., when and where seen. We wrote articles about how to participate in various RMBL activities – the marching band for Crested Butte's 4th of July parade, or submitting a design for the annual T-shirt contest. We listed stories about comings and goings and interesting events and people in Gothic. It was fun; it helped us know each other, it made us feel connected to each other and to Gothic and RMBL."[2]

As people retired and moved on, many others joined with Dorothy and Chris Johnson in assuring the continued publication of *The Gothic News* well into the twenty-first century. Although discontinued after three decades, RMBL continues to compile, print and distribute semi-annual newsletters, in spring and in fall.

Involving the Community

The kind of science being done in Gothic, Colorado at the Rocky Mountain Biological Laboratory has certainly evolved over the years. As previously mentioned, during the initial twenty five years or so the work was mostly observational. Each researcher would choose an individual or category of plant or animal to focus on and record information about it for purposes of observing changes from year to year.[1] Sometimes, a species was newly discovered and recorded for posterity. Some of the species were enshrined in artwork, others dried for archival selection. The investigators who conducted their research in Gothic during those earlier years have been described as "naturalists" by the scientists of the latter half of the twentieth century, who have moved forward in a more "experimental" fashion. This in no way is meant to diminish their credentials, the value of the work they did or the organization they helped to create and build. Then or now, the work done was ***science*** – something that is traditionally held in awe by many outside the field and not infrequently misunderstood.

Gothic may appear as an enclave unto itself, but it would not have existed without interaction with the good people of Crested Butte, Gunnison and beyond. There were cabins to be built, fuel to be found, supplies to be acquired, food to be delivered. Tony's Conoco in Crested Butte was the center for communication with the outside world in an era before cell phones, or even before telephones in general.[2] What did the people of the local community think of the butterfly net-wielding summer residents in Gothic? That is not a question that can be easily answered due to the passage of time; however, there is an awareness exhibited in the archives of RMBL about the need for good relations with the town. The Johnson family members cultivated friendships with townspeople, inviting them to play their accordions at the weekend dances held in Gothic, hiring them to help build cabins or cook in the kitchen, perhaps just getting together for tea, a meal or a chat. Both John and Chris Johnson were very conscious of the fact that Gothic needed Crested Butte, and that Crested Butte was a part of their existence.

As time passed and the little town of Crested Butte began to grow from a mining community into a ski resort and summer playground, the road that cut through Gothic became more frequently traveled. Cars drove through on their way to summer sporting activities such as fishing or hiking.

Bicyclists traversed the road on their way to complete the loop around Schofield Pass and Paradise Divide. People stopped to take photographs of the old cabins, often trampling sensitive research plots in their ignorance of what happened there. Community outreach became inevitable for survival.

Driving along the highway from Crested Butte back to Gothic one day, Mary Ellen "Mel" Harte was chatting with a friend about the community. They picked up a hitchhiker who wanted a ride to Gothic so he could run back to town. She and her friend started discussing the coyotes that had been howling one previous night and the hitchhiker suddenly said, "There are coyotes out there?" That comment motivated Mel to find a way to bring awareness of the research being done in Gothic to the people in the valley. "There are a THOUSAND stories out there and we are just sitting on top of this iceberg of a thousand stories of what is going on in the natural world." It occurred to her that a radio show might be a fun way to convey information. Working with the staff at KBUT, the local Crested Butte radio station, she conceived a weekly summer program and called it "Nature Notes".

For many summers beginning in 1999, Mel invited scientists doing research in Gothic to spend a half hour with her, once a week, to discuss their work. Those interviewed include Bill Calder, discussing hummingbirds; Rosemary Smith explaining burying beetles; Barbara Frase on woodrats; John Harte on bio-diversity, sagebrush and, interestingly, bird watching. Ken Armitage and Dan Blumstein discussed their marmots; Becky Irwin talked about invasive plants; Bobbi Peckarsky shared information about her alpine stream work. Butterflies, invasive plants and insect sex were focused on by Carol Boggs. Scottie Willey talked about salamanders. Locals John Marc and Sue Navy shared information about wild mushrooms and the Gunnison Sage Grouse. David Inouye was there to talk about whatever he was working on in the moment – skunk cabbage, climate changes, drought, wild flowers. Paul Ehrlich and Graham Pyke, among others also sat with her and told of their work. All in all, it was a fun and effective community outreach project.[3]

From that time forward, the scientists at RMBL have engaged in community inclusive events and activities. Mel Harte created a cd of the flora in the area called, "What's that Wildflower?", and has served as a continuing workshop leader for the annual Crested Butte Wildflower Festival. David Inouye frequently writes a column in the local newspaper about something of interest to people locally. Ian Billick hosts dinners at the dining hall in Gothic for supporters and their guests. A summer camp for children has been operating for years. Local supporters Beverly and

Balie Griffith conceived the idea of a "Marmot Club" for people who wanted to support RMBL, an initiative that has grown into an important part of RMBL's fund raising program. Tours are offered free of charge; workshops are available for a small fee. People flock to Gothic to learn about the fascinating work being done there, to learn about it from the people who are actually doing the work. And there is nothing like that stop at the Gothic Visitors Center for a cold drink or ice cream on the way home after a full day of fun in the mountains.

The Rocky Mountain Biological Laboratory reached out to the community, and the community continues to respond.

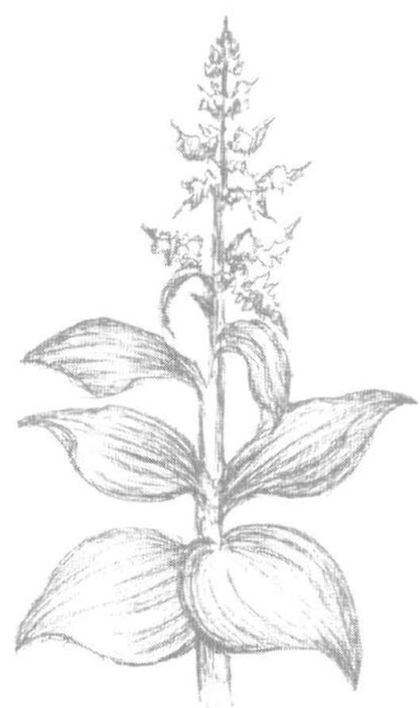

The Mexican Cut Preserve
A Personal Reminiscence, August 2011

The day dawned clear and cloudless, a Colorado August day - this one held no promise of rain, unusual for monsoon season which seems to have come early this year. We donned our hiking clothes, climbed into the Jeep and headed up Washington Gulch road. Keeping promises made, the specific location of this little patch of paradise will not be revealed. Suffice to say, it is in the mountains outside of Crested Butte and generally accessible only by happenstance or with the assistance of a guide from the Rocky Mountain Biological Laboratory. We were fortunate to have an unofficial guide in the person of Doug La Follette, for whom RMBL and all it stands for has become a lifelong commitment and passion.

Doug, no scientist but an environmentalist and great RMBL supporter, had been to the Preserve many times over the years, alone, with friends or as a guide for students. His goal this day was to hike our feet off and give us a glimpse of the research site. For a man in his 70s who spends only about two weeks a year at altitude, he is amazingly spry. It seemed we chased after him for hours. But, it was all in fun.

We parked the car along the road and began the hike with a water crossing – one which took us almost up to our knees and soaked our shoes. It was an unusually late melt this year and the streams were still running fast and deep. Ten minutes later, we were approaching the gate, with its warning signs:

"Property of The Nature Conservancy
MEXICAN CUT PRESERVE
Access by Permission Only
This preserve is managed as a scientific research area by the Rocky Mountain Biological Laboratory. In order to protect the plants, animals and geological features, visitors may not camp, hunt, build fires, remove wood or collect plants and animals on the preserve. No motor vehicles or bicycles allowed."

"Sensitive Research
In Progress
ACCESS BY PERMISSION ONLY!
Absolutely No Dogs!
No Hunting"

The gate and the signs, of course, would stop no one intent on trespassing. But, the nature of the area in which the Preserve is located is such that any persons coming across it would most probably have respect for the environment and treat it accordingly. That, at least, is the hope.

It took a good forty-five minutes to walk the relatively forgiving path to the first cut. The path itself is an old mining road, overgrown with wildflowers of every variety and used by RMBL researchers to access the research area. From the looks of it, there had been no vehicular traffic in months. Nor could one see violating foot prints trampling the wildflowers. All was peaceful and pristine.

After a very short while, we decided that chasing our friend would only serve to minimize our experience and enjoyment of this lovely spot, so we slowed our pace and stopped regularly to enjoy the views. The Maroon Bells looked lovelier than they do from the Aspen side and it was fun to locate our regular hiking trails across the valley. We caught up in time.

Our first stop was to look at an old mining cabin, long collapsed, next to a mine that was large enough to have granted some silver to the miners of old. We talked about those adventurous individuals, climbing into the wilderness with a burro and a pick and hoping to find the proverbial pot of gold – not at the end of the rainbow, but somewhere in the Colorado mountains. Under the spell of those mountains, one can begin to understand the impetus that drove such people farther and farther west, higher and deeper into the hills. Whatever became of them? Disappointment and shattered dreams most likely sent them home after a winter or two. Mother Nature does not take kindly to the disruption of Her domain.

And where are those research ponds, we wondered. Not yet, said Doug. We have places to go before that revelation. So, higher we climbed, bushwhacking all the way as there are no paths to this particular heaven. The fields were alight with color – magenta paintbrush, scarlet gilia, sunflowers, mountain laurel, blue hanging chimes, bluebells, some we could not identify and, of course, the ever beautiful columbine. Little elephants heads stood tall in the wetter areas. Kings crown rimmed the stream and, in patches larger than we'd ever seen them, queen's crown, paler and more variegated than their male counterparts, lusher and bursting with complexity.

We had decided not to attempt to reach the "hidden" lake at the top of the Cut this day. Obligations in town limited our time, so we chose to go to the third level. Along our way, the second level was a welcome stretch of flatter land, skirting small ponds that appeared to be empty of life. We paused to take in the view, the flowers – and our breath – before ascending once again toward the third plateau. Spots along the trek were more like rock climbing than hiking, with their sheer elevations and toe holds. We took our time, both out of desire and necessity, but finally emerged at what seemed to be the top of the world. We could see in almost all directions, our view to the north blocked only by the rest of the mountain and its "hidden" lake. This was the picnic spot, windy and sunny, warm enough, quiet with no sign of the usual marmots or pica - only some bear scat which did not appear to be too fresh. We hoped.

We settled ourselves on flat, comfortable rocks and pulled out our meager snacks. As we munched and talked about the area, the clouds began to drift in. They were white, unthreatening, and we appreciated the brief moments of shade as they passed over. We agreed that, on a "normal" August afternoon, those clouds might foreshadow storms, filled with lightning and thunder – most frightening if they caught you out in the open, unprepared. But this was not to be one of those days. We felt charmed, lucky and privileged.

After our rest, we chose a different direction to descend. We were going to make a loop and, along the way, walk from almost the source of this unique watershed, along the melted snow's path, down to the special research ponds. The flowers had become sparse and less colorful as we had climbed higher and higher; now, they were returning with a vengeance, especially along the banks of the stream that we followed most of the way down. Patches of snow dotted the landscape and ice clung to the banks of the higher lakes. Doug told us that, in all his years of visiting the Cut, he had never seen snow in so many places in August. It was not enough to disrupt even a climb to the summit, but spread randomly throughout the acres, suggesting the past winter and the one to come. We marveled at how, only a few short months ago, this entire area had been covered by blankets of deep snow, as it would be again, perhaps starting in as little as a month. The flora that exist in these fragile ecosystems have such a short while to gestate, grow, blossom and reproduce. Perhaps that is why they seem so proud of their colors and shapes, so filled with strength and endurance. We took care to each walk our own path, so as not to do serious harm along the way.

As we began our descent, we looked wistfully at the path to the lake and the peak. It would not be easy going, even without the patches of snow, but was doable. I learned later that evening, in discussion with a scientist at the Lab, that there is a bronze plaque attached to a rock by the lake, a memorial to the son of the couple who so generously donated funds to The Nature Conservancy to purchase that section of the property.[1] I regretted not having made the effort to get there, but one needs to have something to look forward to next time.

The landscape was marred in spots by the yellow, sulfurous tailings of long abandoned mines. Most of them were small, seemingly just an attempt to taste what lay below the surface. None appeared to be as large as the one we had seen when first entering the Preserve. We took a few moments to peer into several of them, but their secrets were entombed in rock or water.

The way down, after abandoning the wet, slippery stream, became rocky and steep, occasioning the need to climb backwards down the sheer slopes. But, too soon, we came to the final phase of our adventure, the salamander ponds, where research is done annually by scientists at RMBL. There are more than twenty of them spread throughout the area and we did not have the time, or the need, to explore them all. After passing a weather research tower, an anachronistic metal needle amid the green of spruce and aspen, we noticed the comfort facilities of the students and investigators who peopled the research site during the short summer months. These were limited to an outhouse and another old miner's cabin, one wall braced by an angled log to keep it from collapsing sideways. What luxury!

Randomly, we chose a pond to circumnavigate in hopes of spotting the little reptiles that lived there. And, much to our surprise, there they were! - dozens of them, resting on the mud amidst the short green growth. Some chose to nestle into the mud itself, camouflaging themselves from predators such as us. These particular salamanders are especially interesting to the researchers because of the dual nature of their existence. Some remain in the ponds; some metamorphose and move to land. The insects they feed on are also a source for long-term research. The ponds are so shallow that they freeze solid in the winters; fish would be unable to survive. The salamanders cleverly burrow into the mud, where they pass the coldest months and wait for their habitat to thaw come spring.[2]

After a leisurely stroll around several of the ponds, we passed through the researchers' work station, which was already battened down for the winter. No one was in sight. It was time to head back to the car and "civilization". As we crossed over the stream we had followed on our descent, I noticed something flitting in the water. One of the little guys had escaped a pond and was passing time swimming in circles in a small pool created by the rocks. There was nowhere for it to go, but it seemed content in its solitude. It did, to our amusement, react to our shadows and burrowed its head in the mud like an aquatic ostrich. Good luck to it.

Back past the abandoned, collapsed miner's cabin, down the old mining road, around the gate and into the wetlands we proceeded. Feet now dry became wet again as we failed to find shallow water on our crossing back to the Jeep. We paused to look back from whence we came and we could see only a shadow of the miner's road, well hidden by protective wildflowers. They had barely noticed our passing – their time was almost up for the year and they were reveling in the warmth of the sun while it lasted, dancing in the afternoon breeze.

This place is, indeed, special. It was, and is, cherished by RMBL scientists who successfully fought to protect it from development, encouraged The Nature Conservancy to purchase it and preserve it, and who continue to work to unlock the secrets it contains. It must – and will – be kept safe by those who know of it. It was an honor and a privilege to be allowed to experience the beauty, peace and tranquility of this small, pristine patch in our world.

Life on the Edge - It's the Little Things

Just about everyone who has spent time living in Gothic, summer or winter, has a tale to tell of something memorable. What follows are just a few…

On porcupines…

"For several years I had a porcupine living under my cabin. Porcupines sing to each other – I thought this was great! Everyone who knows anything about porcupines seemed to know this, but I did not. The first time I experienced it, I was sitting outside my cabin in the middle of the day. I thought at first that a group of tourists had mistaken the trail by my cabin for the trail to Judd Falls; it sounded like a young girl was coming up the trail, singing under her breath. I got up to say hello and ask where they were going, and there was this porcupine waddling up my path, singing to itself! Then, under my feet I heard a really high pitch singing; another one was under the cabin and the two were singing to each other.

"Also, they like to chew on the wood of the cabin at night. It is so quiet at night [in Gothic] and it sounds like the porcupine must be the size of a dining room table! My porcupine responded to my pounding on the walls to stop. She had a baby under the cabin for three years and I think she was there for four or five years. I don't know if it was the same porcupine, but there was always a little pile of quills I'd collect and give to the young children."

Dr. Barbara Frase

"Someone had brought a dog over from Australia when he moved to Stanford. The dog was supposedly half dingo. It was really a daft dog. It went after a porcupine and the porcupine got the better of it. I remember driving the owner and the dog in the back of the car down to Gunnison to the vet, the dog whimpering all the way. It had quills going from one side of its nose all the way through and out the other side! The dog was eventually okay – and the porcupine was certainly fine."

Dr. Christine Turnbull

"How do you know when it is time to pump out your outhouse? From Harriet Barclay's perspective, it is – if you sit on a porcupine that is sitting on the contents of your outhouse, it's time!"

Dr. Bobbi Peckarsky

"Once when my young son went to the outhouse, he opened the door and there was a porcupine sitting on it! He slammed the door and ran home."

Dr. Clint Chambers, Physician

"One night I got up at about 2:00 am in McLeod cabin. I didn't take a flashlight – it was a starry sky and I like to get out and see the stars. I had to go down a path to the outhouse. I heard a rustle inside, so I went back in and got a flashlight. When I looked, there was a porcupine on the seat, chewing the wood!"

Dorothy Johnson

On catching birds...

Bill Calder had a great sense of humor and was a bit of a practical jokester. Lorene Calder, his wife and partner recalls, "We usually had an assistant or two during our RMBL summers, some more used to the outdoors than others. One summer when we had a mist net strung up near the cabin, Bill wanted to instruct our female assistant on the methods of removing a robin from the net. Robins are notorious for being one of the messiest birds to remove from nets as they frequently void. Bill had pocketed a few strands of cooked pasta, and while he demonstrated the removal techniques, he palmed the pasta and began to eat the strands. The assistant was satisfyingly horrified!"

Lorene Calder

Bill Calder had a special hummingbird feeder outside the Richard's cabin, where he lived. "When a hummingbird perched, he could get its weight automatically from the balance on the feeder. Then, he had on the other side of the cabin a feeder that was inside a cage with a trap door. There was a little string that went in from the window. You could sit and watch for a hummingbird to enter the feeder in the cage, then pull the string and close the trap door. There was the hummingbird to band! Bill used to sit in his cabin and do his research. He would catch hummingbirds for us. We would do experiments where we would train hummingbirds to visit hand-held flowers. Then we could present flowers of different morphologies in any sequence we wanted and see how much pollen was transferred. Bill would capture the birds for me; I would work with them for a little while and then release them."

Dr. Diane Campbell

On emergencies...

"The not funny story was when I was disturbed up at my cabin one time by someone who said that Bob Willey had had a heart attack. This was around 1983. There were very few other people around. We used to have doctors come through in the summers, but the doc wasn't around; he was fishing or something. So, we found a truck with a flatbed and one seat in the cab. We dragged Bob onto the bed of the truck, someone jumped into the cab [to drive] and we shot off down the mountain to Mt. Crested Butte. Bob and I were wrapped around each other, rolling around. He was gasping and grunting and so forth, as you do when you have a heart event. But it turned out okay; we got down there to learn that's all it was, an event; it wasn't a full-blown heart attack."

Dr. Andrew Beattie

On relationships...

"I found, way back then, that being tall and lean and having an Australian accent was the secret to success with the girls. There were lots of girls!"

Dr. Graham Pyke

"A former girlfriend of mine showed up one summer with another guy who was working at RMBL. They camped out up the road for a while. One day, her actual boyfriend arrived to pick her up and take her off to graduate school. The three of us guys all played music and got along pretty well. So, we organized a band for a night of entertainment and called ourselves The Three Boyfriends."

Kevin Donovan, Builder

On mice...

There may not have been hot and cold running water at RMBL, but there have always been "hot and cold running peromyscus"! [deer mice]

Dr. Ken Armitage

"I stayed in the Oh-Be-Joyful cabin before it was remodeled or renovated. It used to be the stables, so it was just drab. The whole bottom floor was awful. One of my roommates was from the Bronx and was really out of her element. She had brought a huge suitcase filled with first aid things, like the antidote for Rocky Mountain spotted fever! But she was studying mammology and set traps under our beds to catch mice. There were plenty of them. During the night I would hear a snap – oh yuck!"

Dr. Bobbi Peckarsky

"The Hull cabin [where the visiting doctors lived] used to have a fireplace – the chimney collapsed a few years ago. When the children were here, they would sleep with their dad [Dr. Clint Chambers] in front of it. I would sleep on the sofa with the dog, a terrier. A mouse would crawl all over me and the dog would just sit and look at it – never bothered to chase it. I looked up and just picked it up and tossed it off me. I was never afraid of mice. It was quite cute – a white footed one."

Mrs. Siva Chambers

On the living accommodations...

"There are two kinds of cabins. If you can pay enough, you get one with hot and cold running water. If you get the cheap one, you get hot and cold running deer mice.

"My feeling is, there was a time when people sort of took pride in being intrepid. My attitude was, okay, been there once; done that. You don't get any science done being intrepid. All the time you spend being intrepid takes away from the reason that you came."

Dr. Ken Armitage

"One year I stayed in Crystal cabin. It was from the 1880s. You could tell that because the walls were papered with newspaper and you could see the dates. It was a bad cabin because of the mice in the walls. You'd hear them all night."

Dr. Clint Chambers, Physician

"It was decidedly primitive back then [50 years ago]. Our cabin was unbelievable – maybe one step up from a tent. It never saw sun, never warmed up. You kept your food in a cage to keep the mice out. The water situation was so bad that were was always a question of intestinal disorders."

Anne Ehrlich

"I never enjoyed roughing it particularly if I was trying to do work. It's one thing to go somewhere, if it's absolutely beautiful, and camp. That's fine.

But if you are trying to do work, camping is a pain. You want to have electricity; you want to be able to keep your notes dry."

Dr. Paul Ehrlich

On music and dancing…

"They used to actually have musicians who played songs. Years ago, Nick Waser played clarinet, Steven Merke played trombone. All it takes is a couple brass instruments and you can actually play something above the clamber of people banging pots."

billy barr

For many years there were folk dances in the dining hall on Friday and Saturday nights. "The old miners from the town would come up and play a couple times a summer. One played accordion and another four-string guitar which I don't think had been tuned since it was manufactured. Chris Johnson overlooked it when they brought liquor with them – the only ones he did that for. It was all very third world – pretty hilarious to see all these hippy students polka around as best we could. It was quite a hoot!"

Kevin Donovan

"In Chris' time there were a lot of things going on. We tried to have things on the weekends so the kids wouldn't have to drive into town. The only places they could go were the bars. We worried about them driving home late, after they'd been imbibing. So Bonnie Inouye would lead folk dancing. The Calders were good about playing music. So we had some fun things to do on Friday and Saturday nights. We used to polka, too."

Dorothy Johnson

"I polka-ed with my husband-to-be. It's the only dance we did together! We went down to Crested Butte and there were polka evenings in the bars, too. That was fun."

Carol Johnson

"I can remember polkas in the dining hall. Everybody would get so sweaty because you really moved around. Of course, we were young and would go all over the place. Then you'd go out on the deck in the cool evenings, cool off, smell that wonderful air, then go back and dance some more. You could dance without a partner, but you really had to have a girl!"

Doug Johnson

"In the early 1980s, Steve Nowicki and I and a few others formed the RMBL marching band. I played clarinet. I had an old, cheap clarinet made of tin or some other metal that I bought in a pawn shop in 1973 or 74 for ten dollars. I still have it – it is my RMBL clarinet. We played in the Fourth of July parade. The RMBL students in those days would show up with nothing on under their [Skunk Cabbage] leaves!"

Nick Waser

On recycling and volunteering…

In the mid-1970s, it was common for almost everyone at RMBL to pitch in to keep the physical plant maintained and functional. "One year, before I left town for the season, I went down the road to change the oil in my car. I knew the Lab used it to paint the buildings, so I brought it back and gave it to Dr. Enders. As I was leaving town, I saw him on the side of the Town Hall. He had pretty copious hair and he had pulled a purple onion sack over his hair to keep it under control. It wouldn't stop anything from going through because it was filled with holes. But there he was, painting my used oil on the side of the Old Town Hall."

Kevin Donovan

On trying to get married in Gothic…

"I was playing music with two friends for Rick and Ro's wedding. As we were playing St. Anne's Reel for the recessional and they were walking away through the field, a long time researcher walked up to me, RIGHT in the middle of the song, RIGHT in the middle of the recessional, and says, 'Uh, this may be bad timing but my toilet's plugged up.' I didn't even lower my eyes to look at her, but I agreed it was bad timing."

Kevin Donovan

On change…

"You know, I volunteered here for twenty years. They gave me a few dollars an hour. This year I had to sign a three-page contract just to do that!!"

Anonymous

On bears…

"We once had a small bear that people described as black, brown, big, small – but it came past the tent where the guys were. It got into the tent and ate all the chocolate chip cookies. Then it went and opened the screen door to the lab, up on the hill above the dining hall. It opened the freezer and ate all the mice bodies and the fish. Then it tried to climb on the roof of the Ehrlich's composting toilet. Paul came out with a little gun and scared it away. The Forest Service tried to trap it but it wasn't interested in their bait, so the students put out chocolate chip cookies, mice bodies and fish. They caught him right away – it was a little brown bear. The Forest Service took him far away and he never found his way back."

Dorothy Johnson

On skunks…

"When I was a small child, possibly around two, I was in the old dining hall with my dad. I had to go to the outhouse. We went to the back door, opened it and promptly got sprayed by a skunk!" — "The smell was so strong - It was just piercing! They got to take the first shower in the new little shower house built for the cook. I finally just buried their clothes."

Carol and Dorothy Johnson

On the dress code…

"In 1949, my first year in Gothic, men were wearing suits and dress shirts and ties. Mildred [Johnson] always dressed in skirts and hose, never work slacks. She wore heels; so did Mrs. Weese."

Dorothy Johnson

"As I was walking across the campus today [July 2012], I thought that the students here now look remarkably the same as they did [in the 1980s] – jeans and a tee shirt, carrying a butterfly net."

Dr. Andy Beattie

On mountain lions…

"One evening, several years ago, in early summer I decided to postpone taking my dog for a walk until I made a phone call. It was just before 8:00 in the evening. I sat down in the rocking chair and the lion was right there, [outside the window]. It came out of the trees and sat down. I dropped the phone and grabbed my camera! I thought what if I had taken the dog for a walk before making my phone call? It could have been a very frightening experience."

Dr. Alison Brody

On Mickey RMBL the Donkey…

Bob and Abbie Enders for several years kept a donkey called Mickey in Gothic. He managed to surprise people with his existence. "The very first day Ward [Watt] was here, he and a couple guys from Yale arrived after dark; they had no idea Mickey existed. Bright and early next morning, on the back side of the cabin with no windows, Mickey let loose a bray. They all went straight through the ceiling!"

Carol Boggs

"One time when people were gathered in the dining hall, someone came running in and, without thinking, cried –'Enders ass is running loose on the hillside again!'" Much laughter ensued.

Anonymous

And So It Is May...

"Although it was late in May and only the western slope was free of snow, the spring of the high country was lying just beneath the cover. In the moist meadows where patches of snow had already vanished, the pointed green shoots of the mountain iris were standing in the sun.

"More wonderful than the iris were the avalanche lilies. They arose and bloomed in the receding snow, and all through Gothic valley their gold heads bobbed against the ice. At night they closed their petals and withstood the freeze and the bitter winds that roared down from the peaks

"The buttercups were also above the ground in the slowly opening alpine meadows, and many of the hardy mountain grasses were coloring the mountain clearings yellow-green.

"The brief, abundant plant life of the high Rockies had begun beneath the snow cover, for it must come to bloom and seed before the autumn of late August and the winter of mid-September. All the alpine life was racing against time. In the plant world, only the hardy perennial species survived the severity of the mountain sun, wind, and ice. The annuals of the lowlands could not mature their seeds in a season so short and they were eliminated at the foot of Gothic Valley.
"As the days passed, more and more life stirred in the high country. Like the golden-mantled ground squirrel, the hibernating mammals awakened slowly. The whistling marmot was one of these." [1]

In such a way did Jean Craighead George describe spring in Gothic in her 1956 children's book, *The Dipper of Copper Creek*. Almost five decades have passed since this was written, but in the eternal way of Nature, not much has changed - except as it has been affected by man and, of course, by long periods of time. Summers are longer, winter comes later. The blooming season for flowers, which for many years was recorded as running from late May until early September is now lasting from late April through the end of September. Scientists at the Rocky Mountain Biological Laboratory have spent decades carefully and regularly charting such information. [2]

We now enter the world of science and the scientists who have made Gothic their summer home for weeks, years and decades. All are biologists, but the areas of study range from mammals to insects to birds to tiny water creatures as well as flowers and plants. Each is fascinating in its own way and has commanded the devoted attention of those who study it, analyze it and draw conclusions about it.

Please Note:

Unless otherwise noted, all the information in the following essays about the scientists and their work at RMBL is extracted from personal interviews with those scientists. The interviews were conducted during the summers of 2011, 2012 and 2013 and 2014.

Marmots and Marmoteers

Dr. Kenneth Armitage *Dr. Daniel Blumstein*

Marmots (genus *Marmota*) are burrowing rodents, a species of ground squirrel that are about the size of a large housecat. They have coarse fur, short legs, short, bushy tails and small ears. There are presently fifteen recognized species of marmots in the northern hemisphere. Native to North America, they live in the mountains of Europe and Eurasia, as well as Canada and the United States. There are lowland marmots, known more popularly as groundhogs or woodchucks in North America and as steppe marmots in Eurasia. In the Colorado Rockies and other mountainous areas of the western United States, yellow-bellied marmots have chosen to make their home. Although they can't know it, these marmots are a special breed, because they are the subject of the longest biological field study project in North America. The Rocky Mountain Biological Laboratory is where it all began – and where it continues.

The time was the early 1960s. A young professor at the University of Kansas, Dr. Kenneth Armitage, was interested in the idea that social behavior could influence population dynamics. Toward that end, he had begun to spend summers studying marmots at the Jackson Hole Biological Station in Grand Teton National Park in northwestern Wyoming. Having become interested in the creatures while working as a ranger naturalist in

Yellowstone National Park, he focused on marmots because they live in groups, they are social and, "heaven forbid, they are active in the daytime so you can see what they are doing". During his fourth year at Jackson Hole he was visited by a friend and professional colleague, Dr. Paul Ehrlich from Stanford University. Ehrlich was on his way to his summer research home in Gothic, Colorado, and suggested to Armitage that he should consider working there, too. Armitage was at a point in his research where he realized he would need to look at multiple marmot colonies, and that the area in Wyoming where he was working was not conducive to the kind of work he wanted to do. It was accessible only with difficulty, requiring hours of daily hiking or high-mountain camping. While this might be exciting for a man without a young family in tow, he decided to seek a more accessible place to study the creatures. So, in 1962 he changed direction and headed for Gothic.

Armitage's first summer in Gothic was when he began his now quite famous marmot study. A project of long duration was not what he originally intended. In fact, he had romantic ideas of working for a few years in Gothic, then traveling the world to study his furry subjects in their other homes – perhaps some time in Alaska studying the hoary marmot, then on to Asia to watch over the Siberian marmot. But it was not to be. As two years stretched into ten and then twelve, and more, he realized that he would make a better, more valuable contribution by staying with the program he had started and "not trying to skim the cream, having fun". A colleague at the National Science Foundation affirmed his decision, telling him he really had no other choice; he had to stay the course. "Scientifically, he was correct," concedes Dr. Armitage. "And so I did."

What, exactly, is this famous marmot project? As mentioned previously, it is the longest running project of its kind in North America, second in the world only to Jane Goodall's chimpanzee study in Gombe Stream National Park in Tanzania. Goodall had begun her study in 1960, two years before Armitage established his research project in Gothic. After more than half a decade of continuous study, Armitage enjoys teasing other scientists who boast about doing something for ten or twenty years, telling them, "Yes, that's a good start".

The role of social behavior in population dynamics was what started Dr. Armitage on the path to his research, but it wasn't long before he began to ask himself what he calls "an old fashioned question: What do marmots have to do to exist where they live?" The question itself is simple and appears to be quite basic, but an answer to it required much more than understanding social behavior. It led him to expand into physiology, particularly water balance and temperature physiology; hibernation physiology; food selectivity, dispersal and, very importantly, the part played by kinship.

In general, marmots live a relatively long time, up to sixteen years as observed by Armitage and his students. Females live together in social groups consisting of mothers and their daughters that may persist for several generations. The female groups are called matrilines and one or more matrilines form a colony. A colony may have from one to three resident males; each male associated with one or more matrilines. Females generally live longer than males. What he found after several years of study was that, "the males were temporary". One would enter the colony and stay until he died, after which time he would be replaced by another immigrant male. Each matriline would have an alpha female who was older than the others. She would choose daughters to remain with her; the sons would be shown the door – or the hole to the burrow, as it were. Some daughters might leave and join another colony, if it were possible. Females from different matrilines rarely interact socially; when they do, the interaction is overwhelmingly antagonistic. Female marmots are not solitary beings; it is the female social group that is the critical group in the marmot world. Males are important only in that they are needed to mate.

Male marmots do not recognize family members as females do. When males encounter each other out in the world, they do not seem to recognize their own offspring or parents, and treat each other as they would a stranger. They are highly competitive in their search for successful residency in a colony. Such residency serves not only to allow the male to breed and reproduce, but also affords him the relative security of life in a colony, rather than on his own in the world. Unattached males are subject to harsh conditions and many dangers, including predation.

Life in the colony is no cake walk for the young females, either. The groups are small and usually consist of two females and generally no more than four. A female may choose to allow a daughter to stay, live in the same environment and perhaps even in the same burrow. However, when a young female comes of reproductive age, at about two years old, the mother may try to suppress her daughter's reproduction. If the mother can prevent it, her daughter may not reproduce until age three or four. If this is the case, why do the young females remain in the colony? In the forty years that Armitage actively worked on his study, no female *ever* invaded a matriline and became a resident. A female "immigrant" can only join a colony when there is a vacancy due to the death of the older animals, who can no longer challenge her. If she cannot join another colony, a marmot is placing herself in mortal danger in the outside world. It turns out that a young female is better off staying home, even with the possibility she may not be able to reproduce in her early years, than she is going out and trying to find someplace else to live.

The system, as identified by Armitage, is actually more flexible than it may appear here. Marmots do live on their own. There is a lot of what is called "phenotypic plasticity". Not to panic here, the phrase simply means that the individual has a lot of choices. "She is not compelled to live in a social group. She can go off on her own. She is not compelled to live in a nice, beautiful environment. She can come out and live under rocks of her choosing and feed in a beautiful meadow nearby. Most matrilines in any year consist of one adult female. All immigrants form a matriline of one; a female who goes off on her own, often in the same colony, forms a matriline of one; and mortality decreases a larger matriline to one." To take the concept farther, there is a great deal of flexibility in what marmots can do. They are not required to eat certain plants, only certain types of plants. However, food supply, reproduction and safety are paramount to the survival of the species. Staying home is almost always safer and more desirable than venturing forth.

Back to the males… Along with identifying female choice in the marmot world, it was important to determine whether females were actively selecting the males with whom they mated, or if the partnerships were based simply on happenstance. Armitage was not able to find any evidence that the females court or discriminate in their choice of partner. There were years when no males were present in some colonies and, therefore, there was no or greatly reduced reproduction. If males arrive too late in the season to mate, they might be accepted into a colony so they can breed when the females come into estrus the following year. But, after years of observation and study, Armitage came to the conclusion that "the female will probably accept any male that's vigorous enough, has lived long enough, to come and say 'here I am'." The female is better off to reproduce when she can, with whatever male is available to her, than to wait and possibly not find a better choice.

In general, males have a lower survival rate after the first year of life than do females. The main reason for that is dispersal. Approximately ninety-eight percent of males disperse. Among the females who disperse, the survival rate is the same as for males; however, as many of the females remain at home, they are safer than their sisters who do not. The average number in a litter is four. Survivorship of the young in the first year is, on average, fifty percent. Some years are better than others, of course.

Since dispersal is a major fact of life for marmots, one of Dr. Armitage's graduate students, Dirk Van Vuren, chose to study it. The males do a lot of moving around, looking for females to associate with. When competing for residency, they are under major stress. They must be on constant alert for predators, especially coyotes, badgers and bears. Young marmots, in particular, are easy prey. The hypothesis about stress in male marmots seems to have been borne out by some laboratory experiments that were done on several that were brought into the lab for the purpose of taking measurements. All of these subjects were known to be in competition with other males in the field. In several instances the subjects died! It was believed that the additional stress from being subjected to captivity, no matter how gentle, well-intended or brief, was more than the creatures could tolerate. They went into some kind of physiological shock and died. This was not part of the plan; the tolerance for mortality due to research by the RMBL crew was zero. The deaths were unexpected, unfortunate and distressing for the investigators.

On a happier note, marmots are not especially picky eaters and are entirely herbivorous. The researchers have not seen any indication of them either preying on other animals or eating carrion. Several kinds of feeding studies have been done over the years, many of them focusing on observation and examination of fecal material. As plants do not have the same digestibility – meaning they don't all break down on the same schedule – it is difficult to determine proportions of ingested foods. Studies have also been done to try to determine if they prefer one plant over another. The conclusion in general is that they like most herbivorous things, and most especially "forbs" – the flowering plants. Having said this, there are particular flowering plants that produce secondary, defensive compounds. Such plants include lupine, fireweed, columbine and delphinium. Marmots will choose to eat the flowers on these plants, but not the entire plant. (Cattle avoid these plants with defensive compounds, also.) They will eat entire plants such as cinquefoil, clover and vetch. Armitage says it is not quite understood why they prefer such plants to grasses, which technically also have flowers, but his interpretation is that grasses don't offer a level of protein high enough for marmots' needs.

Marmot sightings are not rare. Visitors who hike or even drive to the high country may see them running around in the open fields, or playing in the relative safety of rocky areas. They chirp out warnings to fellow marmots when danger approaches. "A marmot will often track a predator, such as a badger, as it moves through a colony. Marmots apparently call to 'inform' the predator that it has been detected. The alarm call also alerts other marmots within hearing distance. By informing the predator that it has been seen, the predator then departs as it now knows that it has no chance for a successful predation and should seek success elsewhere. The advantage for the marmot is that it induces the predator to leave and the marmot can forage or engage in other activities in relative safety."

Interesting to note, although there is only one species of marmot in the mountains around Gothic, they do not all share the same color coats. Some are golden, others more brown, with variations in between. When asked about the discrepancies, which are not related to gender, Armitage responds, "If I could somehow take thirty years off my life, one of the things I'd like to study is coat color". He goes on to say,

"Coat color is very important in thermal regulation. Every place where it is hotter and drier, marmots are lighter in color. Yellow-bellied marmots, same species but a different sub-species, that live over in eastern Oregon, eastern Washington, at rather low elevation, are smaller. They are much lighter colored and they are more heat tolerant than the ones here. So there are physiological fur color adjustments... Is the amount of variation seen [around Gothic] just random? Does it mean anything? The lighter color might be an advantage when it's hot, because you absorb less heat; if you're a dark color, it is an advantage when it's cold, because you absorb more solar radiation. So, does a darker coat have a bit of advantage in the spring, a lighter coat more of an advantage in the summers? It would be hard to work out because you'd have to know what is going on internally as well."

It appears that there is still much research to be done about marmots.

Although it may be hard to believe, and distasteful to contemplate, people do eat marmots. Not the *Marmoteers*, of course! They are devoted to studying and preserving the lives of their subjects. When asked if he had ever tasted marmot meat, Professor Armitage replied, "I am NOT a cannibal!" He does concede, however, that some of the older ranchers and miners he has come across claim that there is nothing tastier than a nice, fat marmot in late summer.

Partly because of the length of marmots' lives and partly, perhaps, due to the nature of the gentle creatures themselves, it is not uncommon for the investigators to tend to identify with them. Dr. Armitage freely admits to going through an emotional change working with marmots. It was unintended; in the beginning, the animals were just a research tool. However, as he began to know them individually, his attitude toward them changed. Consciously working to maintain his objectivity, he also recognized that the animals were becoming an extended part of his family. He trapped and marked "Chia", a female, first as a yearling, then as a two year old, again at four, five and six years old and, finally, as a thirteen year old. He felt as time went on that he knew her, as he did the others he worked with over the years. To illustrate how deeply he felt a connection between the creatures and his own family, he tells of a young marmot he trapped and marked the same year his son was born. That same marmot returned as a yearling, when his son was a one year-old. Each year afterwards, the marmot returned, as did his son, until the year they both would have been nine. The marmot did not return and Armitage thought, "Oh, my gosh! Will my son make it to his ninth birthday?" He did and has made it every year since, but the marmot was never seen again.

Such attachment and devotion to one's research subjects is not universal. But then, marmots are not everyday creatures.

The Rocky Mountain Biological Laboratory is built on private property, belonging to the nonprofit institution. However, the surrounding land is not policed and interference with research projects in general has been an ongoing problem, especially since many of the projects are located offsite on National Forest land. A good rapport and mutually beneficial relationship was established early on with the Forestry Service, as well as with the Fish and Wildlife agents, which has resulted in cooperation to protect the research in general. In regard to the marmots, specific issues surfaced early.

In the early years of the project, individuals with no compunction about killing for sport would all too frequently drive to Gothic and shoot marmots, apparently for fun. This included marked and tagged animals that were part of the project. One animal was found shot in its trap; other traps were found empty but shot up. At the time, marmots were classified as varmints, which meant that there were no restrictions on killing them. But there were restrictions on hunting from the road, so on at least one occasion four young men were arrested for that when their car was reported to authorities. Armitage credits the Fish and Wildlife people with arranging to have the Gothic area closed to activities of this kind.

In the years since, marmots were re-classified as game animals. In line with this, it was necessary to establish a hunting season. As the leading expert in marmots, Dr. Armitage was asked to attend a meeting in Montrose, Colorado, to help with this scheduling. Though not in favor of the process, he searched around in his mind for a way to protect them most when they were young. There was a guideline that disallowed killing of animals until they were independent of their mothers. Knowing that young marmots become independent earlier at lower altitude than at higher altitude, and also that there were known cases where a mother died but the young managed to survive by foraging, he arbitrarily chose a date when official independence could be declared. That date is August 10th and it continues to be the official opening date of marmot season, which lasts until the fifteenth of October. Colorado Small Game Regulations allow for a daily limit of two animals, and a possession limit of four.

Summarizing four decades of work is no easy task. There are, however, generalizations that may be applied to most courses of study. Dr. Armitage put it this way:

"I would say that all organisms are limited by their ability to use resources. If they over exploit those resources, then at some point they are going to pay a high penalty. What I find interesting about the marmots is they apparently do not over exploit their resources. The reason seems to be that before the population builds up big enough to really impact the vegetation seriously and decrease their supply of food, summer is over. They run out of time; they have to go into hibernation. Next year, everything recovers. I think that probably makes marmots different from many other animals that just continue on a year-round basis exploiting.

"They have adapted themselves to live in what I call a harsh environment. They have approximately four months of active season. In those four months, if they are going to reproduce, they have to grow; the young have to start getting bigger, and they all have to put on enough fat so that by the end of the summer they are able to face getting through hibernation. Four months isn't very much time.

"Actually for marmots, time is one of the critical factors. I hope that I can get the idea out there. It's an old idea, that time is a critical factor, but it's not been widely adopted. I think that's partly because people don't study anything long enough. Most studies are three to five years; that's the length of a doctoral dissertation. Or funding runs out. And you don't get the long-term continuity we've been able to get with marmots and the chimpanzee study."

Changing the Guard

Dr. Armitage retired and, in 2001, turned his marmot study over to Dr. Dan Blumstein. After forty years, it was time for a new generation to not only continue the work that had been done, but hopefully to expand on it. Blumstein was a perfect heir.

Dan Blumstein had applied to work with Dr. Armitage as a graduate student, but didn't know at the time that he wanted to study marmots. He ended up going elsewhere and "reconciled with his inner marmot" in a most interesting way. For several years he was paid to lead expeditionary mountain bicycle trips to exotic places. While in northern Pakistan in 1987, he stumbled upon marmots that were interacting with foxes; he actually observed a marmot chasing a fox away. Camping on the Chinese border for a day and half, Dan watched the "huge" marmots he had found there. Returning to work in Pakistan from 1988 through 1993, he decided he was interested in social behavior and, in particular, in play behavior. Why is it that animals play? And, more specifically, what makes grown animals play? What are the costs and benefits of play behavior, and why does it take place around predators?

The University of California at Davis is the place Blumstein chose to pursue his graduate work, developing a research proposal to work with marmots. However, he reached out to Dr. Armitage in Kansas, reminding him that he had turned down an opportunity to work with him and asking if he might still be welcome in Gothic. Would Armitage be willing to teach him how to "marmoteer"? The answer was, of course, yes! Dan spent his second of three post-doctoral fellowships with Armitage at the University of Kansas studying a variety of marmots. During this time, he put in one full summer and a partial summer in Gothic. Having learned the techniques of "marmoteering", Blumstein charged off into the world to pursue his own research.

Blumstein's interests led him in some interesting directions. Working for years around the world, he developed a curiosity about the evolution of language and used marmots as a model system. A post-doctoral grant to run a research group in Australia allowed him to pursue work that interested him about kangaroos, wallabies and birds that evolve in an isolated environment. But he continued to write about marmots. Returning to the States in response to a job offer at UCLA, he again contacted Armitage about working with him in Gothic. As Armitage was getting near to the end of his field days, he encouraged Blumstein to work with him and assume responsibility for the marmot study. "What is remarkable about this is that it is one of the few examples of an intergenerational hand-over of a long term study," says Blumstein. "Those are hard."

Armitage, for his part, has not been able to let go completely. No longer involved in the collection of data, he continues to collaborate with Blumstein in analyzing data and in writing.[1] Once *a Marmoteer*, always a *Marmoteer.*

The new generation has taken and continues to take the research in new directions. Blumstein oversees personality work, where researchers "play with a love drug" and give marmots oxytocin to see if and how that modulates their social behavior. Social studies of marmot social behavior are ongoing, and a recently started project looks at the "dynamics of gene expression", how that is associated with various social and non-social stressors.

Twenty-first century weather extremes have taken their toll on the Gothic, Colorado marmot population. Many of them died during the winter of 2010-2011, following a particularly long, cold winter. More were lost the next winter because of an inadequate snow pack, which is needed to buffer them from the cold. In 2012, the animals emerged in pretty good condition. They had not bred extensively the previous years and so "they didn't suffer the cost of reproducing and then being too thin to reproduce. A marmot's whole life is about being fat". A period of drought may have taken even more of a toll on the animals, but welcomed rain seemed to restore them and many females reproduced and gained weight. Unfortunately, some of the populations and sites that have been a part of the study for seventeen generations, or fifty-one years, have become extinct. But others have been created and scientists are optimistic that they will thrive.

Blumstein has one of those minds that takes a germ of an idea and carries it forward into dimensions he may not ever have anticipated. His work with marmots has allowed him to expand into unexpected and fascinating areas of study. One has to do with sound. In his words,

"I listen to my inner marmot. Everyone should. I live in Los Angeles, so there are not a lot of marmots there. But marmots scream – I study alarm calls and alarm communication. Pups sometimes scream, so I became really fascinated with screams. Darwin knew what screams were. He didn't know what marmots were, but many other animals scream. So what are screams and what are cries? It turns out that acoustically they are very distinctive and different from normal alarm calls. Screams are produced by many vertebrates that produce sound by blowing air across something that vibrates – vocal cords in mammals, the syrinx in birds – and are probably the outcome of overblowing vocal production systems until they produce noisy and chaotic sounds. If you blow a trumpet too hard, it gets noisy. Similarly, if you force air too quickly through your vocal cords, the resulting sounds sound noisy. That noise is a defining characteristic of screams and cries for help. So, queue to films now. I don't meet a lot of marmots in Hollywood, but I meet a number of film directors and composers. So I've started a whole series of collaborations with musicians where we are trying to understand what humans perceive as the sounds of fear. Turns out that the ideas that emerged from my studies of marmots in Colorado have informed my studies of music and human emotions."

Blumstein explains that his marmot study is a single species study. It deals with only marmots. However, while he is studying marmots, he is asking broader questions that hopefully are more generally applicable to understanding the diversity of behavior we see in Nature. Thus, he hopes that the answers to some of these questions apply to species outside his study. An example has to do with the way rodents breed. The females have two uterine horns, like a "v", and along each horn are little pods and little placenta. Inside a pod, an embryonic female might be surrounded by one brother, two brothers or no brothers. An embryonic male might be surrounded by one sister, two sisters or no sisters. Hormones leak through the placentas and this influences the development of other individuals. Thus, a female surrounded by two brothers is exposed early on to slightly higher levels of testosterone. This defeminizes her. "Anogenital distance (I love this word!)" is what is measured to determine if an individual is a "manly girl" or a "girly man". But why does this matter?

For lay people this may mean nothing, but Blumstein eagerly continues with his tale. First of all, the consequences for the individual rodents are profound. They are less likely to survive the first year, less likely to breed when they become mature, more likely to disperse (which is often fatal for females) and more likely to engage in rough kinds of behaviors at all ages. And here we return to the significance of the broad questions.

"Now, let's think about plastics and estrogen-mimicing chemicals in the environment, and modern chemistry and all the junk we have floating around in our food chain, that we are eating. Why are girls going into estrus at age ten or eleven when they never used to? Why are people becoming obese? The point is that slight changes early on in the pre-natal environment, the hormonal environment, have profound effects throughout an individual's life. These may even extend into the individual's offspring. What is shocking from what we've learned about marmots is that slight differences in natural variations of these hormones can have profound consequences, so think about the really bad consequences that might be resulting from human-caused pollutants. We should be really careful about what junk we put into the environment. This is NATURAL variation, not chemical. There may be no safe limit to these things if they are endocrine disrupters."

This is all by way of saying that, we are what we eat and breathe, and what we do to our environment comes back to bite us.

Dan Blumstein's research takes him farther afield than the university classroom or the fields of Gothic. Government agencies are interested in some of his conclusions, or at least in the direction of his work.

When asked if the project will continue, Blumstein's response is, "Letting this study go is a war crime! The insights you get just become more valuable and more valuable over time. RMBL gets great undergraduates to work with me. It provides a rich intellectual atmosphere, where I've certainly grown in a hell of a lot of other dimensions, and I really like that. Yeah, I'm coming back here! Of course, I'm coming back here!"[2]

And as long as he returns, the study goes on, into its sixth decade and eighteenth generation of marmots.

Beautiful Butterflies and their Kin

Dr. Charles L. Remington (1922-2007)

Butterfly. The word itself conjures up images in one's imagination of color, design, gentleness, freedom of spirit, lightness of being. It brings a smile to the mind, if not to the lips. Chasing butterflies is akin to following one's dreams. They flit through the sky, light briefly on an endless number of brightly flowering plants, working naturally as pollinators and unknowingly as a calming balm to stressed human minds. Should one land on you, it imparts a sense of magic. They are too beautiful, too delicate to be real.

Real they are, however, and the man responsible for so much of what is known in the world today about Nature's exquisite fluttering palettes is Dr. Charles Remington, "Father of Modern American Lepidoptery".

Charles Remington was introduced to *Lepidoptera* – butterflies and moths – at a young age by his father, an avid collector. The journey of discovery took him to many places in the United States and eventually to graduate school at Harvard, where he pursued a PhD in biology. Prior to that time, however, he spent his World War II military years as a medical entomologist, working in the Pacific Malaria Survey Unit. In this capacity, he helped to identify the sources of and stop the progress of insect-borne disease epidemics such as dengue fever in the Philippines and typhus in Japan. These and other experiences helped to broaden his interest in insects in general.

After completing his doctoral work at Harvard, Remington was asked to teach for a year at Yale University, filling in for an absent professor. He stayed for forty-four years, retiring from the Department of Ecology and Evolutionary Biology in 1992. His interests were broad within the fields of ecology, genetics and evolution. He was a dynamo, "so bursting with ideas that he couldn't possibly settle down to one narrow thing" explains his former wife, Jeanne, who worked at his side for many years. Because of his bounding curiosity, he left much of the writing about his work to others, including graduate students and Jeanne. He did, however, publish some of his work.

While still at Harvard in 1947, Remington and fellow student Henry K. Clench established "The Lepidopterists' Society". Their goal was to "promote the scientifically sound and progressive study of *Lepidoptera*" by publishing a newsletter and facilitating the exchange of ideas, information and specimens among professionals and amateurs. Within a month of the distribution of the first newsletter, there were 220 members. The organization now claims members in more than sixty countries. [www.lepsoc.org]

In addition to The Lepidopterists' Society, Remington helped to found The Connecticut Entomological Society and, along with Dr. Paul Ehrlich, the organization originally called 'Zero Population Growth (ZPG)', now renamed 'Population Connection'.

Charles Remington's own words in a privately published autobiography titled, "A Passion for Insects", give us a greater sense of who he was and what he accomplished…

"Only serendipity can fully explain my varied and dramatic journey in the skin of an evolutionist. I've never shed my wide-eyed wonder at what I've witnessed on that journey…

"I have indeed been at the right turning point at the right watershed moment. Could I have planned, for example, my World War II medical service to encounter and help defeat two serious insect-borne epidemics or have found a way to treat suffering servicemen who had been bitten by a foot-long centipede? How could I have expected to discover, in my Louisiana army camp, quantities of the tiny mosquito that uniquely proved only to take the blood of amphibians? I later happened to do my doctoral work at Harvard at the same time as the great writer [and lepidopterist] Valdimir Nabokov… [who] was to become a great friend. I was in a unique position to study the Periodical Cicada upon its massive seventeen-year hatch in 1996, to try to introduce new colonies, and later to stumble onto a colony of rare giant cicada killer-wasps. In collaboration with Dr. Gary Bernard, I was able to be a principal part of the groundbreaking research on the color vision of butterflies, [which] came to pass at the Rocky Mountain Biological Laboratory, a facility I helped to expand. My studies of speleobiology were improved by the sudden opportunity provided by the U.S. National Park Service when they opened new and virgin passageways in Mammoth Cave and Carlsbad Caverns. The chance crisis of an infestation of book beetles at Yale University's Beinecke Rare Book Library caught me when I was ready to solve it. Finally, the gallery of brilliant theoreticians and insect collectors I've had the good fortune to know is an admirable group that has, individually and collectively, shaped the worlds of entomology and evolutionary ecology."

Whew. Remington certainly bounced from one interest to another, responding to opportunities he was able to recognize and act upon. His research extended from *Lepidoptera* to silverfish, cicadas and cockroaches to insects in caves and insects on islands. What, then, is his connection to the Rocky Mountain Biological Laboratory?

Charles Remington's friendship with scientists such as Colin Pittendrigh at Princeton University and Edward Novitisky, at the University of Missouri, led him to his first introduction to RMBL in 1954. Prior to this, he had taken his family to spend summers at a science lab in Nederland, Colorado, run by the University of Colorado. Invited to give a paper at RMBL in 1954, Remington returned with his family in 1955, arranged to have a cabin built and continued to visit periodically until 1988. The research he conducted in Gothic and its environs was mainly focused on *Lepidoptera,* a perfect topic because of the vast diversity of butterfly species in the area. Dedicated to the spirit of RMBL, Remington arranged for visits by prominent scientists, including modern evolutionary biologist Theodosius Dobzhansky and structural geneticist A.H. Sturtevant, helping in this way to raise the profile of the small, independent field station.

In 1988, Remington spent his final summer in Gothic working with Yale University's Dr. Gary Bernard on a study of color vision and behavior in butterflies. A generous grant from the National Science Foundation allowed Bernard to move his "extensive physiological equipment" from New Haven, Connecticut to Gothic, Colorado, a prime site for both the various butterfly populations and the necessary laboratory work. Having studied butterflies in Gothic for many years, Remington was uniquely qualified to provide information on the local species. Their stated goals were to discover what the color spectrum is in vision of several families; if there are receptor zones in the compound eye; if there are gender differences; and, in what ways do females use color vision to identify from within the large green world of the meadows those small number of acceptable, nurturing plants on which they could lay their eggs.

The researchers concluded, among other things, that in many, if not all, butterflies color perception is similar to that in the human eye, and possibly more acute in the shades of red. There tend to be three zones of "ommatidial receptors" in the hemispheres of the eye that are similar in males and females. It is possible that "there may be major differences functioning in courtship", concentrating on the zones where the butterflies look at each other. And, females may see differences in greens reflected from the acceptable plants, but before laying their eggs they use their forelegs to puncture the skin of the plants and taste what lies beneath it to make sure the plant is acceptable.

Pursuing the question of color vision in the animal world, Remington says in his autobiography,

"What animals have color vision? Many or most do not. Among insects, we suspect that most insects do not, but orders other than Lepidoptera should be investigated. The simple photoreceptors of juveniles… may not see colors. Most mammals do not, including dogs and cattle; it is amusing that the ancient sport of bullfighting uses red cloth to excite and attract the bull, but bulls apparently cannot see red. [There are] mammal exceptions. Most birds (but not owls) presumably discriminate colors. Many (most?) lizards see colors; supposedly snakes, turtles and crocodilians do not. We do not know about amphibians; perhaps all are color blind." In other words, there is much that continues to be unknown about color vision in the animal kingdom.

Perhaps one of the most significant contributions made by Charles Remington to RMBL was to bring other scientists and students to work in Gothic. Among these investigators are Dr. Paul Ehrlich and Dr. Ward Watt. . Dr. Ehrlich has been in summer residence at Gothic for more than fifty years, studying butterflies even as his interests have broadened to human population and environmental concerns. Dr. Watt began his relationship with Remington when he was still in high school, corresponding with him and eventually moving to Yale to complete his education. His work at RMBL has "consistently focused on comparative evolutionary genetics and physiology, using *Colias* as a test system – the point not being to study butterflies, but to use butterflies to study evolutionary processes". According to Watt, "my intellectual relationship with Charles has not been to follow in his footsteps, but to have mutual interest in parallel, non-overlapping scientific questions which he and I asked separately."

Ward Watt likes to share a story about the importance of Charles Remington in the study of butterflies. In the summer of 1998, The Lepidopterists' Society held a conference in Mt. Crested Butte and Remington was made honorary president. At one point during a session, the people in the room were asked to stand up if they had been a student of Charles Remington. About half did. Then those still sitting were asked to stand if they had been a student of Paul Ehrlich. The rest stood. How many academic generations of *Lepidopterists* were in that room? It is hard to know, but the number is significant.

Charles Remington passed away on May 31, 2007, but his legacy remains.

The following poem, "The Butterflies' Fad" by Ella Wheeler Wilcox, was one of Charles Remington's favorites. He would recite it from memory to willing listeners.

The Butterflies' Fad

By Ella Wheeler Wilcox

I happened one night in my travels
To stray into Butterfly Vale,
Where my wondering eyes beheld butterflies
With wings that were wide as a sail.
They lived in such houses of grandeur,
Their days were successions of joys,
And the very fast fad these butterflies had
Was making collections of boys.

There were boys of all sizes and ages
Pinned up on their walls. When I said
'Twas a terrible sight to see boys in that plight
I was answered: "*Oh well, they are dead.*
We catch them alive, but we kill them
With ether – a very nice way,

Just look at this fellow, his hair is so yellow,
And his eyes such a beautiful gray.
"Then there is a droll little darkey
As black as the clay at our feet,
He sets off that blondie that is pinned just beyond
In a way most artistic and neat.
And now let me show you the latest, -
A specimen really select,
A boy with a head that is carroty red
And a face that is funnily specked.

"We cannot decide where to place him,
Those spots bar him out of each class,
We think him a treasure to study at leisure,
And analyze under a glass."
I seemed to grow cold as I listened
To the words that those butterflies spoke;
With fear overcome, I was speechless and dumb,
And then with a start – I awoke!

Source: Entomological News, Vol. 6, September 1895, p. 226. Available on google books, www.books.google.com.

Heeere's... Paul!

Dr. Paul R. Ehrlich *Anne H. Ehrlich*

If a superstar title should exist in the world of biological sciences, Paul Ehrlich must be able to claim it. His name is certainly more familiar to the general public than most scientists of his ilk. And, it doesn't hurt that he was a guest on the Tonight Show with Johnny Carson more than twenty times! Garrulous and opinionated and occasionally inelegant, he exhibits a level of comfort with his positions about life in general and the scientific conclusions his personal analyses have helped to reach. He is not your stereotypical scientist. And many people would be surprised that he began his career as a *lepidopterist,* someone who studies butterflies.

Paul R. Ehrlich had a natural interest in butterflies as a child growing up in New Jersey and was concerned about their dwindling populations. "I had watched the butterfly sites that I worked on as a kid disappear under housing developments. I couldn't raise butterflies anymore because they had sprayed DDT." In college, at the University of Pennsylvania, he began reading about the effects of DDT on the environment, and as a graduate student at the University of Kansas he worked on a project dealing with the evolution of DDT resistance in fruit flies. Such was the beginning of his life-long interest in all things environmental.

When still a teenager Ehrlich met the prominent bee researcher,

Charles D. Michener, who was curating the butterfly collection at the American Museum of Natural History. Michener directed Ehrlich to The Lepidopterists' Society, which had recently been founded by Charles Remington and which he joined. Through that association he met not only Remington, but also his student, Ward Watt. After graduating college, Ehrlich sought Michener's assistance in being accepted for graduate study at the University of Kansas (KU), where Michener then taught, and in obtaining a research assistantship with Professor Robert Sokal, a prominent biologist there. At KU, Ehrlich met Ken Armitage, who would be on his dissertation committee, along with Michener and Sokal. With all these connections, could the Rocky Mountain Biological Laboratory not be part of Ehrlich's future?

It was in graduate school that Ehrlich met the woman who would become not just his wife and partner in life, but his life partner in work. A one-time art major, Anne Howland began her life and career with Paul illustrating his doctoral dissertation. She has since co-authored many of his books, including the most generally well-known, ***The Population Bomb*** (though her name was left off that one because of publisher's restrictions). They moved to the Department of Biology at Stanford University in 1959, where they continue to live and work.

The Ehrlichs first visited Gothic, Colorado with their young daughter in 1960 at the urging of Charles Remington, who arranged for Paul and another scientist to teach a course in biostatistics. That first summer was a bit of an eye opener, the primitive living conditions of the facility contrasting with the excellence of the research opportunities and the natural beauty of the area. Fully cognizant of the realities of life in Gothic, the family decided they most definitely wanted to return again and again, but did not want to live in such primitive conditions. So, they contracted with RMBL for a renewable twenty-year lease on a piece of property of their choosing and built a cabin. They have lived in that cabin and conducted research from it every year since 1961, with the exception of two sabbatical summers.

Paul continued his work with butterflies, both at Gothic and at Stanford, where he was on the faculty. "It was clear to me that for the kind of work I wanted to do on butterflies [Gothic] was an ideal place. You have high mountain stuff that comes down from the arctic; you have desert stuff basically that comes in from the Great Basin; you have eastern elements; you have things that come up from Texas during the summers and invade here. It's a place where you can step out of your door and start doing research right on the spot. Plus, it is one of the most beautiful places we'd ever seen. And, since we had a full-scale research program going on at

Stanford in the spring, this offered a good way to add a second field season."

One of the early projects Paul and Anne collaborated on, and that was the major focus of their work for several years, was to try and determine how often female butterflies mate – and if they even do so more than once. At the time of insemination, males leave a plug in the female in order to prevent another male from seeding their territory. It was discovered that in some cases the plug will last a lifetime, making the female essentially monogamous, "whether she wants to be or not!" In most instances, however they found that the plug does wear off after a while and another male can gain access.

For all the time they spent in the field, Anne claims that she was "not good out there". Her job was to dissect female butterflies and study their reproductive tracks – using butterflies obtained for her by others. Paul and friends and even their young daughter would catch the butterflies and give them to her to work with. She just didn't have the knack. "I once almost caught a butterfly and at the crucial moment the net came off the handle!" But, she was very good at collecting larvae and managed to do some field work in Gothic doing that. It was all about teamwork.

During the course of his more than fifty summers at RMBL, Paul Ehrlich authored over forty papers based upon his work there, some of them in the field of coevolution.[1] Coevolution is defined as changes that occur in the genetic composition of one species in response to changes in the genetic composition of another species – or, how two species evolve in response to each other. He was involved in several long term studies of the structure, dynamics, and genetics of natural butterfly populations, especially checkerspot butterflies, in the genus *Euphydryas.*

Along the way early in his career, Dr. Ehrlich became interested in human population dynamics. Asked to give a talk about it at the Commonwealth Club in San Francisco, he became widely known for his ardent positions on the effects of overpopulation. The immediate result of this talk was the book, ***The Population Bomb***, published in 1968. This popular and controversial book was followed up in 1970 with a "serious" textbook, ***Population Resources Environment.***

The controversy surrounding Ehrlich's dire conclusions about population growth may be in part responsible for his celebrity. He has refused to bow to critics who point out that his predictions of widespread famine were wrong. Instead, he continues to pursue his cause. In an interview on National Public Radio in September 2012, Ehrlich defended his thesis:

"The problem is, of course, ***The Population Bomb*** was much too optimistic a book... for example, we knew the climate was going to change, but we had no idea how fast or how bad it would be, which we are learning dramatically now. Of course, agriculture is our source of food... agriculture is staggering and yet it is responsible for about twenty percent of the greenhouse gases going into the atmosphere. So we are in the terrible position of, the more we try and push agriculture, the more we change the climate; the more we change the climate, the tougher it is on agriculture. The situation is indeed dire. It hasn't really changed dramatically since 1968. We were already hitting our support systems hard in those days, and now, of course, we are hitting them much harder. We missed a lot of things in ***The Population Bomb*** like ozone depletion, the resource wars that are now basically starting up and what was happening to biodiversity, which is our life support system, basically. If I wrote it again, it would not be any more cheerful."[2]

Many other books have followed and the Ehrlichs have been involved in founding and supporting many institutions concerned with issues of population dynamics and climate change, including the Center for Conservation Biology[3] and the Millennium Alliance for Humanity and the Biosphere (MAHB).[4]

At the end of the day, however – or more accurately, each summer, at the end of the academic year — Paul and Anne Ehrlich return to their little cabin in little Gothic, Colorado. Paul has brought many a graduate student to Gothic. He and Anne keep in touch with many of them. Some of those graduate students continue their associations with RMBL. The academic family continues to grow. And RMBL continues to benefit. The Paul R. Ehrlich Research Fellowship has been established at RMBL to provide financial support to a promising scientist to work at the Rocky Mountain Biological Laboratory, "whose work takes advantage of the knowledge accumulated in the Lab's 80+ years".

Of all the people who have passed through the Rocky Mountain Biological Laboratory, Paul and Anne Ehrlich are among the first, if not the best examples, of how their work there helped to shape their perceptions of the world around them. They share those perceptions with their fellow scientists each summer, many of whom are engaged in work to reveal the realities of climate change. Those realities, in turn, have helped to shape the research that followed and that is yet to come.

Johnny Carson would be proud.

Those Evolving Butterflies

Dr. Ward Watt

An early fascination with genetic variation in butterflies led a young Ward Watt to contact Dr. Charles Remington when still in high school. Remington had helped to organize the budding Lepidopterists' Society while pursuing his graduate work at Harvard University in the late 1940s. A correspondence ensued between the professor and the youth resulting in Watt's enrollment at Yale University, where Remington had gone to teach, for undergraduate and then graduate work. A life-long friendship and professional association began, with Watt arriving in Gothic in 1962, at the behest of Remington. Beginning his field work that summer, he has returned every summer since, save three. Even after joining the faculty at Stanford University on the opposite side of the continent from Yale, Watt's relationship with Remington flourished, based on their shared enthusiasm for studying evolution by using butterflies as "test systems".

While at Yale, Ward spent many hours sorting through the butterfly collections in Yale's Peabody Museum of Natural History. An observation he made there launched his career. He noticed differences among certain butterflies of the genus *Colias*. Those that lived in colder climates had darker hind wing undersides than those living in warmer climates, thus absorbing more sunlight and heating up more effectively to allow flight. Basing himself at RMBL for summer field work, Watt did his dissertation research about, and began his life's work on, the evolution of metabolism and physiology as exemplified in *Colias*.

Glycolysis is the metabolic pathway by which animals convert food to energy. Enzymes are proteins that control chemical reactions. PGI (phosphoglucose isomerase) is an enzyme of glycolysis that is found in all animals, indeed in nearly all cellular life. Ward's research found that *Colias* butterflies show much genetic variability in the PGI enzyme, which differ in function over a wide range of temperatures. It turned out that butterflies having more efficient versions of the PGI enzyme were able to fly more at optimal temperature, as well as over a broader range of temperatures. But, butterflies that had less efficient versions of PGI were sometimes able to survive longer when exposed to heat stress, thus extending the PGI enzyme's variability.

Connecting these findings to evolution, it turns out that the length of time a butterfly flies is related to the PGI variants' efficiency. Longer flying time allows males more time to find and court females; it allows females more time to find suitable sites for laying eggs. Thus, there is an important connection between how long a butterfly flies and the number of offspring it produces. In the course of this study, Ward was able to analyze the evolutionary success of variants in a single gene.

The growth of modern technology has greatly helped to facilitate the course of this work, which is rather ironic due to the historically less-than-high-tech nature of the Rocky Mountain Biological Laboratory's facilities. However, investigators such as Ward Watt have gradually introduced new technology to RMBL and encouraged the modernization of its facilities. Beginning with a simple observation about color differences in butterflies, Watt has spent his career studying evolution. "He has sequenced genes, resolved protein structure, examined metabolic pathways and watched what butterflies do in nature."[1]

One of the longest continuing research investigators in Gothic, Ward Watt is a singular source of information about the work that has been done and continues to be done at the Rocky Mountain Biological Laboratory. He is also a living history source of information about the community, sharing recollections of the past and recounting stories that inspire eager new scientists.[2]

Tropical Transitions

Dr. Carol Boggs

Carol Boggs first visited Gothic and the Rocky Mountain Biological Laboratory in 1978, shortly before she and Ward Watt were married. A tropical biologist by training, the focus of her work at the University of Texas, Austin had been tropical butterflies. Both she and Ward were a bit dubious as to how Carol would adjust to Gothic, a most non-tropical environment, but she immersed herself in work there. She chose to focus her research at first on the Mormon fritillary butterfly, or *Speyeria mormonia,* a relatively close relation to her tropical insects, but has gone on to work also with other species. She has not missed a summer since 1978, even if her stay had to be a short one.

Insect mating practices may not be a frequent topic for dinner table conversation, but it is one that has been studied by more than one generation of lepidopterist at RMBL.[1] Boggs is no exception. During the courtship process, the males of many species of insects offer the females what is known as a "nuptial gift". Depending on the species, that gift may take the shape of a dead item of prey, or might be simply something to attract the female. Oftentimes, it includes nutrients which the female uses to nurture herself and the eggs during the process of fertilization. With butterflies, the gift is a "spermatophore".[2]

A spermatophore is a round object that looks somewhat like a pearl. It contains not only sperm, but nutrients to sustain the female during her life. A male butterfly deposits it inside the female during mating. The female saves the sperm in a "storage organ" until ready to fertilize, using a tooth-like object on her body to rupture the spermatophore. The spermatophore and its nutrients are absorbed internally by the female, who uses it for egg production and general personal maintenance.

Protection of the sperm and guarding its genetic interests is of concern to the male butterfly. From the viewpoint of the first male to mate with a female, it is not desirable for eggs fertilized by the sperm of a second male to be nurtured by the nutrients in the spermatophore of the first male. Therefore, a process called "mate guarding" exists. In butterflies, the spermatophore not only looks like a pearl, but is quite hard like one. Females have a nerve that is stretched by the presence of the spermatophore and gives her a sensation of "fullness", the idea being that if she feels full she won't want to mate again. However, once she absorbs the spermatophore, there is little reason for her not to go looking for another mate. Studies show that although females mate at least once, they may mate more often. For the female, there appears to be no argument against - and many in favor of - multiple mating.

Carol's work with the Mormon fritillary butterfly has been ongoing. An area in which she continues to be actively interested has to do with "how environmental variation in food availability, or in available flight time, influences reproduction, survival, dispersal, metabolic rate and other life history traits". Part of the research both, in the laboratory and in the field, has focused on the nectar on which butterflies feed.[3] In the Upper East River valley the butterflies exhibit a strong preference for nectar from the *Erigeron*, or fleabane daisy. What she and her assistants have learned was surprising to them, as it was not in keeping with common wisdom prior to the experiments. They have shown that "if nectar is quantitatively reduced, then the butterflies lay fewer eggs but have equivalent lifespans to control butterflies that were fed as much as they wanted to eat". Other, more specific, data have been accumulated, but there continued to be a nagging question about causes of variations in the size of the butterfly populations from year to year. Mentioning this in passing to fellow RMBL researcher, David Inouye, he offered her thirty years-worth of records about the blooming of the fleabane daisy. Combining her data sets with David's, Carol was able to link the annual population size of Mormon fritillary butterflies in the Gothic area to the number of flowers per butterfly in the previous year, which in turn was linked to the date of snowmelt, which is recorded annually by billy barr.

One might not expect a lepidopterist to be terribly interested in invasive plants, but exactly the opposite is true. After all, butterflies (and other insects) rely on plants for their existence. Plants are a source of food, but they also provide a platform on which to deposit eggs and a nutritional smorgasbord for the developing caterpillars. In 1975, Dr. Frances Chew discovered a problem with butterflies and an invasive mustard plant in the Gunnison Valley. She did not follow up on it at the time but, twenty years later, Carol Boggs decided to pursue it, with Francie's permission and occasional collaboration.

There are native species of mustard plants in the Gunnison Valley, but by the 1880s an invasive European species, *Thlaspi arvense,* was introduced, presumably by the miners who arrived in the area at that time. A species of white butterfly, *Pieris macdunnoughii,* lays eggs on the native mustard plant. When the time comes to do this, the female butterfly uses little hooks on her front legs to scratch the surface of the plant and tap it so she can taste the chemicals. If the chemicals are acceptable, the plant is used as the depository. Chemicals in the non-native plant are acceptable to the butterfly like those in the native plants. The trouble begins when the caterpillars hatch and try to use the plant for food. Either they taste it and don't like it, or they don't try to taste it at all for some reason. Though there may be other viable food sources nearby, these little guys are the size of a pinhead, so it is not possible for them to migrate elsewhere. The end result is that they die - a disastrous outcome for the butterfly.

What Boggs began looking at was the impact of the non-native mustard species on the population and evolutionary dynamics of this white butterfly. The project is focused specifically on three questions: Is there variation among females in the preference for laying eggs on the mustard? Is there variation among larvae in terms of whether or not they can survive on the mustard? And, where may evolution take them in the Gunnison Valley? Is it possible for the butterfly to evolve so that caterpillars can survive if they hatch on the non-native plant or can the butterfly evolve to stay away from those plants?

The experiment in plant preference involves bringing female butterflies into the laboratory and testing to see what plants they prefer. This is done by offering them the non-native and a variety of native plants. Once laid, the eggs on each plant are counted. The offspring are reared in the lab, and the same process is repeated. The similarity in plant preference between mothers, offspring and grand-offspring is used to determine whether plant preference is genetically determined.

Having done that, the researchers took their work into the field, examining the natural egg distribution. The environmental conditions under which the plant was growing were taken into account, since they influence the chemicals in the plant to which butterflies respond. Is a host plant growing in the sun or the shade? Is it in the process of seeding? Does it have lots of water or is it in a drought zone? Are the plants growing closely together in groups, or are they isolated? Do these things make a difference in a female's choice of host plant?

Results showed that in the Gothic town site, approximately three percent of white butterfly eggs are laid on the non-native plants. These eggs have absolutely no chance of survival. A loss of three percent of the offspring may not sound like much, but it results in a very large selection force on the butterflies.

And what about the plants? The caterpillars are not eating them, so are their growth and distribution affected in any way? This is unlikely. These invasive mustard plants thrive in dry, well-drained areas and particularly like to settle in spaces that have been cleared for construction or other reasons. They are even used for anchoring areas of unstable soil. If there are no butterflies in the area where they proliferate, there are no known negative impacts. But in places such as the Gunnison Valley where the butterflies might mistake them for the native plants, the negative impact of their presence is significant.

Yet another project in Dr. Boggs' repertoire involves a checkerspot butterfly, *Euphydryas gillettii*. A population of this species was intentionally introduced to the Gothic area, then dropped to extremely low numbers, but then suddenly exploded after about twenty-five generations. The reasons for this are the bases of research being conducted in Gothic and in the laboratory.[4]

Boggs' involvement in what is a long-term experiment involving the Gillette's checkerspot began in 2002. Walking to her lab one day, she spotted an unfamiliar butterfly. Wondering if it might be the elusive insect that Paul Ehrlich had for decades asked everyone to keep watch for, she captured it and called Ehrlich. It was, and he asked her to meet him at the study site. Not having been part of this study when it originated, her somewhat baffled response was, "Great, but where's the study site?" Directions in hand, she and her research assistant, Ipek Kulahci, met Paul at the site that had been established by him and Cheri Holdren many years earlier. What they found was that the population of the butterflies there had "exploded to roughly 3,000 adults in that original area". As with all long-term studies, there is a back story...

In 1977, Dr. Paul Ehrlich and Ph.D. candidate Cheri Holdren began discussing genetic questions about the long-distance dispersal ability of a particular checkerspot butterfly. They were also interested in testing the idea that "empty niches exist – that more species can be packed into a given area than currently live there". Paul Ehrlich's concern for population issues was not limited to humans! Receiving permission from the RMBL Board of Trustees, they introduced into an area of the RMBL property about ten thousand eggs and larvae from a population of butterflies in the Granite Creek area of Wyoming - a population that would later become extinct, a victim of drought. By making this introduction into a new environment, the investigators were hoping to test not only the "empty-niche" concept, but also what happens with "translocations", or the movement of species into new areas (which is one possible response to climate change). Very finicky about the habitat in which it lives, the Gillette's checkerspot did establish successfully in Gothic.[5] The population of adult butterflies, eggs and caterpillars was carefully monitored throughout the 1980s by Ehrlich and his assistants, during which time "the adult population size varied between twenty and two hundred individuals in an area the size of two football fields".[6]

Update to 2002, and Carol Boggs' discovery led the investigators to realize that the population had then "exploded to roughly three thousand adults in that original area!" Further investigation over the course of that summer revealed that the Gillette's checkerspot had "also expanded to cover the entire Gothic townsite and down the East River. Over the next two years, the population contracted to along the East River, in patches of suitable habitat, but with the primary population numbers still in the original introduction site". Since that time, the adult population of these butterflies has fluctuated dramatically, from about one hundred to as many as ten thousand in the original introduction site. A satellite population was established in 2007, about half a mile away from that original site. No new populations have been spotted since that time.

Why is this significant? Why is any research significant? There may be a specific question that begins a project, or it may be a shot in the dark. But any study, especially a long-term one, can result in answers to questions that were never initially asked. Ehrlich and Holdren's original project emanated from a discussion about the interpretation of genetic data from another checkerspot species. What resulted was the fortuitous "rescue" of a species that became extinct in its original habitat (although it still exists in a wider range), as well as the opportunity to explore some fundamental questions in biology. "The first, and most obvious, question is, what is causing these huge fluctuations in numbers, and why is the population fluctuating at higher numbers now than in the 1980s?" It is known that part of the answer is climate change. Other factors are involved, however, such as the role of parasites and egg predators, both of which are under ongoing investigation.

The next set of questions is related to genetics. The population was subject to what is known as "founder effect" – that is, it was established with only a "sampling of the genetic variation present in populations" in its native habitat range, hundreds of miles distant. Since the time of establishment of the colony in Gothic, there has been "no migration of new individuals, with different genetic make-up, into the population". Since this population has also been subjected to several "bottlenecks", or periods of time when the population size fell to as few as twenty adults, some genetic variants were lost due to chance – depending on which individuals survived and which did not. This Gothic population of Gillette's checkerspot, with its severely reduced genetic diversity compared to populations in the native range, gives investigators the opportunity "to test some long-standing hypotheses about how much of the genome's variation is maintained by selection and how much can be lost due to chance events. This has been tested in the lab with fruit flies, but almost never in the wild".

Once again, a multi-generational study is taking RMBL scientists in directions they did not necessarily think to look.

An enthusiastic resident of Gothic, Carol Boggs likes to tell of how the area has changed over the years since her arrival in 1978. "When I came, Elk Avenue was paved but none of the side roads [in Crested Butte] were. The road to Gothic was only one lane wide. There were no telephones for another nine years. I'll never forget the first time the phone rang in my cabin – I just about jumped out of my skin! You had to go to Crested Butte, or Mt. Crested Butte, where there were payphones to make a phone call. Or, you could use Tony's [in Crested Butte]. Tony was RMBL's lifeline. There was a radio phone, because Bob and Scottie Willey were in the Civil Air Patrol. For the most part, though, if you saw the marshal come up here, either someone had a job offer or somebody had died. There were no fences; there were no cars on the road at all. You knew everybody on the town site because there were so few of them."

No one seems to really bemoan the fact that the Rocky Mountain

Biological Laboratory, and with it, little Gothic, have grown and are growing. But it certainly is fun to remember the way it was.

Oh, and yes, Carol Boggs has continued to work on tropical butterflies off and on, but her heart remains in Gothic.

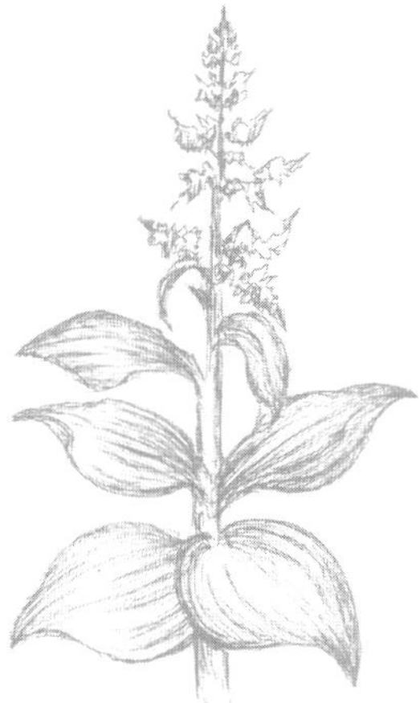

Harte-to-Harte

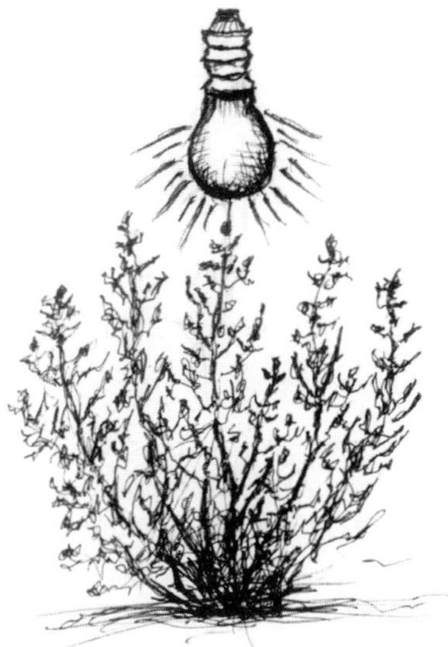

Dr. John Harte   Dr. Mary Ellen Harte

"Global Warming". "Climate Change".

Who hasn't heard one or both of these catch phrases for the odd behavior of the weather and the climate in the past several decades? Is the world getting warmer? Is the climate really changing? Does it snow more, rain less, feel hotter or colder; does the sun shine brighter or are the days grayer? Sunshine and gray days have to do with weather. Climate, however, is much more complex. It is not just about temperature, but is associated with storms, drought, flooding, heatwaves, wildfires – the list goes on and on. Since the 1980s, climate change has been a major focus of work done at the Rocky Mountain Biological Laboratory by Dr. John Harte.

A degree in physics is not what you generally find among the scientists at the Rocky Mountain Biological Laboratory, but John Harte was trained as a physicist at Harvard and the University of Wisconsin-Madison. RMBL is a *biological* field station; why ever would he choose to work there?

The year was 1977 and John was on the faculty of the University of California, Berkley. A friend of Paul Ehrlich's, he agreed to spend the first of what would be three summers in Gothic co-teaching a course with Paul and several others on "Biology and the Human Predicament". The course was a broad survey of the environmental dilemmas as they were perceived in the mid-1970s and hosted some major guest speakers on environmental issues. John led students on field trips, acquainting them with the basic science underlying the various environmental crises. Along the way, he visited the Mexican Cut – and it was love at first sight. "It is like a natural laboratory; all those ponds!" On his first visit there, looking down from the summit of Galena Mountain to the ponds, he posed the question to himself, "Where are the nutrients coming from?" This would be the beginning of one of his most well-known projects leading to significant results concerning acid rain.

Mary Ellen Harte, known generally as "Mel", is a zoologist whose field of specialization was marine science. She joined her husband, John, in Gothic starting in 1982 and also joined him in his growing devotion to studying, proving and publicizing the reality of climate change and the science behind it. Beginning in 1999, Mel created and hosted a program called "Nature Notes" on the local Crested Butte radio station, KBUT. Many investigators from RMBL were interviewed during the multiple year run of the program, the focus of which was to educate the local community about the work being done in Gothic. Mel also became involved with the local Wildflower Festival as part of her efforts to educate people about the flowers. Since 2002, she has been documenting the local flora using digital photography. The photographs essentially comprise a digital herbarium at RMBL and are available online.

First, to acid rain…Considering that question of where the nutrients come from on Galena and other mountains, Harte continued with his mental deliberations. "Everything is flowing downhill. We were watching water cascading down the hill and rain and snowmelt washing through the soils. There must to be sources of nutrients because if you are up high on a mountain and everything is flowing down carrying nitrogen, phosphorus with it, you have to restore those nutrients. So, where do they come from?" The question was one that no one at RMBL was investigating, so John decided to pursue it. Initially, he collected rain in buckets and measured the amount of nitrogen and phosphorous and other essential plant nutrients in the rainwater. Not expecting to see much, he was surprised to discover the water had quite a bit of nitrogen in it. "In fact, it was in the form of nitric acid; acid rain." Acid rain is a conveyor of nitrogen, but in a form that can do harm. So in 1980, Harte returned to Gothic for the summer to begin a major study of acid deposition at the Mexican Cut. For eight years, every rain storm and every snow storm were systematically measured, with the help of billy barr; data on the chemistry of the rain and the snow was collected. The results were significant.

Unlike much of the research done at RMBL, John Harte's work is often tied to what he calls "policy relevance". In other words, there have been political repercussions. This is true with both the acid rain study and his work on climate change. His involvement with science as a part of politics began with the acid rain investigations.

For the sake of comparison, rain and snow were collected not only at the Cut, but also in Gothic. Careful documentation showed that there was definitely "acid rain", with some of the rain storms having a ph level of 4.2, "which is like lemonade – pretty acidic". The logical next question was, how was it affecting the ponds at the Cut? Study showed that with the annual snowmelt, there was a "sort of flush of acid into the ponds", bringing their ph level to 4.7 or 4.8. (The lower the ph number, the higher the level of acid.) By mid-summer, the levels would be back to 6.0 or 7.0, "perfectly fine for the creatures". A ph level of below 5.0 is of real concern to biologists, because many organisms do not do well at that level. There was concern for the salamanders in the ponds, as well as fish in other waterways. The salamanders in the Mexican Cut ponds became the next major focal subject in this study.

In 1981, an annual census of salamanders in all of the ponds was begun. Visible numbers of the creatures were counted every week, all through the summers, with the "young of the year and over-one year-olds" counted separately. Naturally, as with any census, it is possible that some animals were missed, but starting in 1981 until 1987, the number of adult population dropped continuously, from one year to the next. As the salamanders can live for ten years, this meant that some were dying and young were not replacing them. "In fact, for 1982, 83, 84, 85 and 86, we saw NO young of the year. Nothing was hatching. We started putting two and two together, realizing that it might be the acidity."

The next step was to harvest eggs of the same species from some ponds lower down in the valley that are buffered by limestone and do not suffer from acid pulses. The eggs were taken to the Cut and placed in isolated containers that were floated in the ponds, keeping them at the same temperature as the pond water. Different levels of acidity were added to each of the containers, ranging from ph 6.0 to ph 4.8. It was found that once the acidity dropped below 5.0, the eggs would not hatch, but above that level there was almost one hundred percent hatching. It was also discovered that there is a small window during the gestation period when the egg is sensitive to acid, which lasts approximately from day five to day ten. Exposure to acid during that five day period causes the eggs simply to wither and die. Exposure before day five or after day ten has no impact.

Now the problem was to determine the source of the acidity. There were several, as it turned out. One was the Douglas Copper Smelter in Arizona; others were coal-fired power plants in the Four Corners region and some smaller plants in the southwest. Winds from the southwestern United States conveyed the acidity to the area of study.

Working at the Mexican Cut was no easy task. Since the investigators wanted to be there before the snow melted, they would arrive late in the spring. John and a couple students "would ski up to the Mexican Cut. We would get up at four in the morning, we would get to the Cut at about four in the afternoon – it was about a twelve hour haul. Sometimes we had to go way around Emerald Lake because the road to the lake was snowy, narrow and dangerous; you couldn't really tell where there was a road. The road was closed above Mt. Crested Butte, so it was all on skis. We'd get up to the Cut exhausted and crash in an old miner's cabin. To get into the cabin we had to dig down; it was completely covered by snow. In 1984, we lugged a kerosene or propane stove up there. We would dig down to the upstairs window and crawl in, sleep there, get up the next morning and spend two days gathering samples. We'd punch a hole in the ice on the ponds and sample the water underneath. We carried food and water, but also kept supplies up there – we had all kinds of emergency supplies. Sometimes, my students would stay up there for weeks at a time. I'd come down after a couple days, recuperate and head back up again. I was beginning to feel I could not keep up the schedule. One of the motivations for locating the climate experiment only 300 yards from our cabin was that it was close".

On to the politics... The Clean Air Act had been passed in 1970, but a movement was underway to amend and strengthen it. A congressman from Wyoming named Dick Cheney decided to hold a hearing in Jackson Hole, hoping to discredit the pending legislation. It was not believed that acid rain was present in the western part of the United States. Presenters from the Environmental Protection Agency tendered studies showing they had sampled hundreds of bodies of water in the west. These samples were collected through the use of helicopters, which would not have been able to access many high mountain areas in May or June; therefore, the sampling was done in late summer.

John Harte was invited to present the evidence that "acid rain was an issue in the west and it was harming a very sensitive population of creatures that were important enough that The Nature Conservancy bought the Mexican Cut". He and Bob Yuhnke from the Environmental Defense Fund in Colorado presented John's data from the study. During the course of the hearings, they were able to establish that the data from the EPA was gathered in late summer, not in early summer when the acid pulse would course through the mountains. By showing that the acid level in the Mexican Cut ponds was normal in late summer, they were able to discredit the data in the EPA studies based on the timing of the collection. When challenged that their data related to only one small area, a unique place, they were able to show that the source of the acid rain itself was well outside the territory of the ponds, and even outside Colorado – Arizona and other areas of the American southwest. The hearings were halted. The Clean Air Act of 1990 would be passed.

The final year of the salamander census was 1988. The Douglas Smelter had moved by that time, the power plants were already cutting back on emissions and, most importantly for the salamanders, it was one of the driest years on record with very little snow. As a result, snowmelt came early and with it the acid pulse – which was finished by the time the salamanders gestated in June. The census that year counted more than one thousand "young of the year". Observing this, Harte ended his study – but it wasn't the end of his work at RMBL. A new concern had arisen – what happens if low snow pack results in ponds drying up due to insufficient water? "Why would ponds dry up? Why would snow melt early? Bingo – climate change!"

These considered observations led Harte to a pivotal study that began in 1990 and has been in operation day and night, summers and winters, ever since. It is the longest running study of climate change in the world and the longest running manipulated experiment at RMBL.

The experiment itself seems quite simple at first viewing. John singled out ten plots on the ridge of the research meadow at Gothic, marking their boundaries with string. Each plot measures three meters by ten meters, or approximately ten feet by thirty-three feet. Half of the plots are left to Nature; half are fitted with special heaters that warm them about four degrees Fahrenheit day and night, year round. One hundred eighty probes are located in the soil, eighteen per plot, and readings are taken every two hours of the temperature and moisture. That data is transmitted to collection boxes, or data lockers, adjacent to the plots and harvested by researchers once a week or so, even in winters when caretakers must ski up the hill to the boxes. Stepping inside the lines of the plots is not allowed; catwalks are set up if it is necessary to access the inner portions. Every summer, the vegetation in each plot is enumerated and cataloged.

Information compiled from the plot data is fascinating and certainly attention-grabbing. The manipulated plots are being warmed in such a way as to mimic what climate is anticipated to be in 2050. Changes were projected for these plots, but the changes occurring in the unheated ambient, or control, plots are interesting. These plots are showing such things as a slow loss of soil carbon, which is being released into the air as carbon dioxide, which creates more warming. "This extra amount of carbon dioxide, coming from a quarter century long warming trend, is adding to the carbon dioxide built up in the atmosphere. This is what is called positive feedback, which is a bad thing." Carbon dioxide causes warming.

Another, more general trend is the change in snowmelt. The records maintained by billy barr, who measures snowmelt every year in Gothic, show an overall trend of earlier and earlier snowmelt which impacts the blooming of wildflowers. This earlier melt, combined with decreased carbon in the soil, increased carbon dioxide in the air and greenhouse gasses is resulting in an increase in the growth of sagebrush in both sets of plots. The bottom line is, sage brush is taking over where there used to be flowering plants.

The increase in sagebrush may not turn out to be a bad thing. "In the long run, it may actually reverse and create a negative feedback to climate change." What this means is that the "flush" of carbon dioxide in the air from the loss of flowering plants might be countered eventually by sagebrush gradually building a soil higher in carbon as it establishes itself. This remains to be seen.

The politics of climate change is in the news on a near daily basis. It is debated by people across the spectrum of science, religion and government. Once more into the breach and armed with his scientific data, Professor Harte leads a charge in favor of changes that must take place if the world is to prevent ecological disaster. He and Mel campaign rigorously in favor of limiting reliance on fossil fuels and working toward a world of "clean" energy. The data from his warming plot study is carefully documented and seemingly irrefutable. It has spawned a multitude of other studies, including 32 journal articles and nine Ph.D. theses. People have come from all over the world to build upon this study. Someone from North Carolina did a study on how ants have responded to the heated versus the ambient plots. A local investigator has been looking at the fungi in the soil in the plots and how the grasses are affected by the warming. Someone else has begun a project on the microbes that live in the plants in both sets of plots. An individual from the Forest Service became interested in the toxicity of delphinium in the heated versus the ambient plots. As delphinium is toxic to cattle, an increase in toxicity could be disastrous to the cattle industry.

Perhaps one of the most exciting studies being conducted using data from the Harte study has to do with evolution. One of John's graduate students, Anne Marie Panetta, has been focusing on the question of whether one of the flowering plants in the plots, rock jasmine or *Androsace septentrionalis*, has actually undergone natural selection in the past twenty plus years. Most of the flowering plants living in the plots are very long-lived. An Erigeron, the common fleabane daisy, is one example of a long-lived perennial. Another is the *Claytonia lanceolata*, or spring beauty, which might live for thirty years or more. "So, in the course of this experiment, these long-lived plants may still be in the same generation that was here in 1990 when we started - with one exception." That exception is the rock jasmine, which is an annual. It will have gone through a new generation nearly every year since the study began, which is long enough for investigators to see actual evolution at work.

Over time, the rock jasmines in the heated plots have come to produce one-third to one-tenth as many flowers as those in the control plots. As "flowering success is a measure of reproductive potential, this means that it has become less 'fit'." Fitness in evolution is the "capacity to hand down your genes to the next generation". This little flower is found in the areas around Gothic at a variety of altitudes – as high as 13,200 feet and as low as about 7,500 feet, proving that it can survive a large range of climate fluctuations. Why, then, is it being so impacted by the heaters? Looking at the traits of the plants – the size of their leaves, their height, etc. – it was noted that they differ according to location. Using this information to carefully look at the plants surviving in the heated plots, it was noted that their traits are more similar to the plants living at the lower altitude, "less like those in the control plots and much different than those at higher altitude". What this means is that we are seeing natural selection in process. The plants are evolving to cope with the effect of the heaters, and this has occurred in two decades. This is good news for the little annuals, but bodes very poorly for the long-lived perennials which cannot evolve so rapidly.

After twenty years of support from the National Science Foundation, the warming study was denied further funding. The very fact that the money was there for so long was quite unusual; the NSF has many applications to consider and generally only funds a project for two or three funding cycles. The project is in danger of being shut down, which would result in the loss of continuing data about climate change. Fortunately, several of Harte's students and others working at RMBL have been able to piece together funds from their grants to keep the heaters running. Harte is hopeful that the Rocky Mountain Biological Laboratory itself will take over the study as an RMBL facility. This would keep it running and make it available to outside users who wish to take advantage of the range of data in their own projects.

As is so often the case with scientists who make Gothic their summer home and choice of field research station, there are stories of commitment to the place that transcend the role as scientist. So it is with John Harte. After The Nature Conservancy had purchased the Mexican Cut lands, John was asked by them to take major donors on tours and talk about his work up there, which he did for about eight summers. One couple wanted to make a substantial donation to the Conservancy in memory of their son, who had died in an avalanche. This donation was for the purchase of the Galena Lake portion of the land, but there was a catch. The couple wanted a large brass plaque engraved in memory of their son to be placed by the lake, in a permanent fashion. The plaque itself weighed about eighty pounds; the glue to hold it fast to a rock weighed another twenty-five pounds and the tools to get the job done added to that. Scottie Willey was involved in these negotiations, but was not strong enough to carry such weight. John agreed to haul this 110+ pound burden on his back up to the lake, where he chiseled out a flat spot on which to glue the plaque. It remains there today, very near the shoreline. "In return", says John, "we got Galena purchased."

The political discussion about climate change is ongoing. But a visit to Gothic, Colorado and a tour of work being done there – or even a conversation with a scientist at RMBL about the work – offers one an opportunity to observe proof in action.[1]

A World of Wildflowers

Dr. David Inouye

Phenology: *A field of ecology that studies the seasonal timing of events in the biological world.*

It is not common for a scientist to discover the Rocky Mountain Biological Laboratory through a non-scientist spouse or partner. It is usually the other way round. But in David Inouye's case, his wife Bonnie visited Gothic three years before he did. She spent three weeks there in 1968 as an *au pair* for the grandchildren of Abigail Enders, wife of zoologist and former RMBL director Robert Enders, whom she knew as a student at Swarthmore College. She conveyed her enthusiasm about the place to her future husband, David, and in 1971 they spent part of the summer there. David took courses and Bonnie worked in the dining hall. He began graduate school at the University of North Carolina, Chapel Hill, that fall. His professional career was spent at the University of Maryland, from where he retired as Professor of Biology in 2014.

An interest in insect pollinators was what initially tied Inouye to RMBL. His first summer in Gothic, 1971, he took courses in mammology with Dr. Charles Handley, curator of mammals at the Smithsonian Institution National Museum of Natural History, and physiological ecology with Dr. Bill Calder from the University of Arizona. In 1972 he began graduate research on hummingbird territorial behavior, independently of and also in collaboration with Bill Calder. He decided to choose a dissertation topic that would allow him to do his field research in Gothic and spent four summers there after switching to work on bumble bees, the topic of his 1976 Ph.D. dissertation. Eventually, he transitioned his research primarily to wildflowers and spent decades on projects concerning them. He continued over the years to work with bumble bees and hummingbirds, as well as "a variety of projects, just by virtue of being here and keeping my eyes open for things that might be interesting to work on. One of the great things about working in a field station is the opportunity to interact with and collaborate with a variety of people."

Inouye was named President of the Ecological Society of America in 2014, forty-one years after the start of his major long-term study on the phenology and flowering of wildflowers in Gothic. That study began as a collaborative effort with a group of graduate student, Nick Waser and Australian biologist Graham Pyke; friends Ruth Wagner and Suzy Calder; and Bonnie. Within a few years the other collaborators had the data they wanted and only David continued to collect data. In all, there are thirty original plots, dating to the beginning of the study in 1973, as well as another set of experimental plots initiated in 2010.

The study was not, at first, intended to be long term. "There were several of us back in 1973 who were interested in wildflowers. We decided that we would each pick a habitat and set up some two-by-two meter plots. At the end of that summer, we had some really interesting information, such as how many species occur in a two-by-two meter plot, how long they stay in bloom and how much they overlap. It seemed that if we did it one more year, we'd get some more information, such as how variable it would be from year to year, so we did it a second year. At the end of that time, each of us had the information we needed." Graham Pyke decided to move on from this particular study, so David assumed responsibility for his plots and continued the work. Graham's plots were along the roadside as you enter Gothic from Crested Butte; David's original plots were along Trail 401, a dry, rocky meadow. He decided it would be interesting to compare the dry, rocky meadow plots with those in the wet meadow.

The original thirty plots, at differing elevations, have been attended to every other day during the research season, from May until September, since 1973. One flower of primary focus is the Aspen Sunflower (*Helianthella quinquevervis)*, also known at the navigator sunflower because it usually faces east, the direction of the rising sun, and for that reason can be used to navigate over open fields. All plants (about 120 different species) are counted every other day to see when each flowers, how long it stays in bloom and how many flowers are produced. These records are now the most detailed long-term in the world on the timing of plant flowering. This project was begun before anyone thought about climate change, and now provides a valuable record for studying how phenology and abundance of flowering are changing. For example, it documents that the entire community of flowers is in bloom for one full month longer than it was in 1973.

The timing of all this flowering is relative to the date of snowmelt, which has been recorded since the early 1970s by billy barr, who monitors a five-by-five meter plot in Gothic all winter. The "official" date of snowmelt each year is the day that billy sees bare ground. (It may not sound very scientific, but it is consistent and now a valuable long-term record itself.) That date changes each year but, since the time these records began to be maintained, snowmelt has come on average earlier and earlier. In 1970, it was June 1st; in 2014, it was May 15th – two full weeks earlier in a forty-four year span of time!

Tending these plots was time consuming, but for many years Inouye was able to manage it while continuing to do research on other projects. Then, after about twenty-five or thirty years, the project became a major focus at RMBL. The National Science Foundation provided funds to continue the research. Money became available to bring in post-doctoral assistants. In 2014, the NSF grant was renewed for five years, so the project will continue for at least that long, possibly being assumed by David and Bonnie Inouye's son and daughter-in-law, both biologists.

In 2010, some experimental plots were established. Some of these plots were located in the dry area around Trail 401. These plots were watered. Other plots were set up with Plexiglas frames to warm them. There are a total of twenty experimental plots, and they differ in a critical way from the original plots, which were subjected only to observation and recording of the observed data. There was no alteration or manipulation of those plots to create an "experiment". Information was tracked as it occurred naturally. As such, the original project was very much in keeping with the work done in the early years of the Rocky Mountain Biological Laboratory, before the 1950s.

The new plots were created as a manipulated experiment. Amy Iler, a post-doctoral investigator who has worked with Inouye, explains: "These [new] plots were set up with warming frames that prevent radiative heat loss from the plots. There are little devices set in the plots that measure the temperature and the amount of light every half hour or so. These get warmed by about half a degree Celsius during the days and by about two degrees at night. What we are seeing with climate change is that night time temperatures are increasing faster than the daily maximums. We set these out right after the snow melts and one of our main interests is figuring out how important snow melt is relative to the air temperature in terms of when the plants come into bloom." This experiment differs from the warming plots set up by John Harte because the warming is passive; there is no electricity used and the plots are not warmed all year long, but only during the research season.

An extraordinarily energetic individual, David Inouye's scientific curiosity extends well beyond the sunflowers in his plots. He has never lost his interest in bees and hummingbirds. There are about a dozen species of bumble bees in the Gothic area. They settle into different parts of the valley according to altitude. A long-tongued bumble bee called the *Bombus appositus,* lives in the Gothic area. Another long-tongued species is common at higher altitudes and yet another prefers lower altitudes. During the course of the past decade, the lower altitude bumble bee has been seen in the Gothic area. A relatively new study being done at RMBL is looking at bumble bee movement; it is showing that the bees are shifting in altitude, moving higher, presumably in response to climate change.

As for hummingbirds, their close connection to wildflowers and the blooming schedule is important; hummingbirds pollinate wildflowers and wildflowers nurture hummingbirds. Each spring, the birds arrive in time to take advantage of the nectar in the blooming wildflowers. A question that is now being investigated is whether the hummingbird migration patterns will mimic the changes in the wildflower blooming patterns. Data compiled and maintained by billy barr over decades indicates that these changes are not occurring at the same rate. Therefore, this is a question that will continue to be monitored. How will the change in flowering patterns impact the hummingbirds?

In addition to this, Inouye is focusing his interest on where hummingbirds go at night. The females stay with their nests in the lower parts of the valley, but it is known from Bill Calder's observations that the males move up in altitude for the night. So Inouye has been experimenting with attaching tiny radio transmitters to male subjects to try to track their movements. Using eyelash glue (of all things!), he attaches the transmitters to the birds' backs. One male was tracked more than one thousand feet up Gothic Mountain before the transmitter fell off. (It took more than two hours to locate the expensive little transmitter.) Future plans are to refine this study and also to attach transmitters to females to locate their nesting sites. Consider the size of a hummingbird and then consider the size of the transmitter small enough to be attached to it. Modern technology certainly plays a role in future experimentation.

A common flower in the Gothic area is the fleabane daisy, or *Erigeron speciosus*. Carol Boggs, who has been studying butterflies in Gothic since 1978, noticed that this particular flower is a major nectar resource for her butterflies. A casual comment to David Inouye alerted him to her interest; he was able to supply her with thirty years of data about the abundance of these flowers. Using this data, Boggs was able to show why her butterfly population was quite variable from year to year. The butterflies she studies rely on the fleabane daisy for about ninety percent of their nectar. Carol and David's combined data sets showed that snow on the ground on a set date in the spring affects the number of flowers available that summer; and, the number of flowers available per butterfly affects the population growth from one year to the next. So, in years of late snowmelt, the flower population is reduced and the population of butterflies suffers. In fact, "about eighty-five percent of the variation in butterfly population is determined by the ratio of butterflies to flowers the year before".[1]

Aspen sunflowers are used as host plants by "seed predators". Some of the sunflowers develop brown spots in the middle of their heads. When broken open, little black pupating flies are visible. This species of fly lays its eggs in the flower heads and the maggots then eat some of the developing seeds, pupating inside the flower head. Some of the pupae hatch out, but some of them are eaten by parasitized wasps. "When we put these flowers' heads in a cup and see what hatches out of them, we get a mixture of flies and wasps. It turns out that almost all of the plants in the sunflower family, including the Aspen sunflower, have these kinds of flies in them." Collaborating with his son on a project looking for how many kinds of flies use the flower heads in this way, David has learned that of the six kinds of related seed-eating flies in the Gothic area, only two use the Aspen sunflower and others use different flowers. "There are a lot of interesting questions we can ask about how predators of these seed predators are protecting these flowers."

The Aspen sunflower is host to an entire community of animals and interacts in many different ways with its natural surroundings. In addition to the seed predators and predatory wasps that rely on them for food, flies and bumble bees collect nectar and pollen from them, and deer love to eat the flower heads. They also attract ants, which can help to protect the flower heads from the seed predators. A study about ants and the sunflowers begun by Inouye has been running for thirty-five years. "It takes quite a few years to figure out all these single stories, to be able to understand what is happening." One study begets another study, which may beget many more studies.

And then there is the Monument Plant, *Frasera Speciosa*. This fascinating plant is also known as miner's candle and green gentian. Prior to the 1970s, the plant was identified as a biennial, developing its rosette one year and flowering the next. Anyone who has hiked in the Colorado Rockies is familiar with the unusual pattern of "mass flowering" this plant engages in. Some years there are only few to be seen; other years the fields and meadows appear to be filled with them. David Inouye and a friend from the University of Kansas decided to take a look at these plants and their flowering pattern, or lack thereof, and in 1973 began tagging individual plants. "Quickly, it became obvious that this is not a biennial plant. At this altitude [9,500], I now know that they can flower as early as twenty years – because I planted some seeds and I waited. Of the seeds that I planted thirty-three years ago, a lot of the plants have not yet flowered!" Some of the plants David has been following at higher altitude are eighty to one hundred years old before they flower and die.

These fascinating plants flower only once in their life and generate, on average, about six hundred flowers. These flowers make a lot of seeds. They have only one chance in their lifetime to attract pollinators, so they produce all these flowers and make them easy for the pollinators to reach and harvest the nectar. But what triggers the mass flowerings? After thirty-five years, a series of unsuccessful experiments and continual observation, David was able to obtain only eight data points (years of actual flowering). He could then ask, "What is the relationship between the environment and flowering?" The answer seems to be that a wet July or August will trigger a flowering event four years later. These plants certainly have a great deal of patience.

An annual count of the monument plants has been a part of this study. It is done very scientifically by Dr. Inouye, who travels up from the ski area to Emerald Lake and counts all the flowering plants he can see from the road. One year, he went to count the plants in a particular yard that was included in his annual enumeration. There were NO plants to be seen. When inquiring of the owner about it, he was told, yes, they were so big that year they were blocking the view. So, they cut down about sixty flowering plants! A conversation about those plants and how old they were ensued, in hopes that this would not happen again.

False skunk cabbage, *Veratrum,* is another infrequently flowering local plant, blooming even less frequently than the monument plant. In the four decades Inouye has been in Gothic, he has observed only three "good" flowering years. These plants are well known to Fourth of July parade-goers in Crested Butte, because the leaves are used to make costumes for students at Gothic who march in the parade. Some of them look like the Jolly Green

Giant; others are barely clad.

The *Veratrum* root is interesting in that it rots away on one end while growing on the other. Inouye says it is easy to calculate the age of the living root segment. They don't flower very often but are "clonal". When they do flower, they put up one to five new shoots around the flower stalk. About twenty or thirty years later, the root will have grown past the point of connection with the new shoots. Although genetically identical, the plants are no longer physically connected. Bizarre stuff! He has identified some clones that are fifteen meters across. As they grow approximately a quarter inch a year, "to be that size and grow so little a year means the clones are five or six hundred years old. They are beginning to rival the age of some of the aspen clones, which are probably thousands of years old up here."

Amy Iler became interested in the flowering pattern of the skunk cabbage. Using some data collected from the phenology plots, she attempted to discover what triggers the flowering. The answer seems to be weather; if there is a relatively cool summer, the plants flower two years later. "So what happens if the climate is warming and the summers aren't cool anymore? These plants won't be triggered into flowering – and they won't be cloning. We are curious and a little worried about the long term fate of these plants given climate change."

A fascinating anecdote about the false skunk cabbage has to do with sheep. Very few animals will eat this plant. In Europe, where a variety of it grows, gloves are used to handle it. An odd event was occurring among some grazing sheep in the western United States that took the help of the USDA to figure out. Shepherds in Utah were having problems with cyclopic lambs – lambs being born with only one eye in the middle of their foreheads. It was discovered eventually that if a pregnant sheep eats *Veratrum* on the 18th day of its pregnancy, its offspring will be born cyclopic. David Inouye can go on and on about the work he and others have done at RMBL. His enthusiasm propels him continually forward in his research projects. He and Bonnie have spent several winters in Gothic throughout their forty-plus years there, and plan to continue returning there and working at RMBL as long as possible. There is real concern about the future of the environment, in Gothic and elsewhere.

"I love these meadows," says David. "I like to work up here. One of the results of John Harte's research is that the warmer, somewhat dryer climate we may see here in the future favors sagebrush over wildflowers. There is some concern that fifty years from now we may be having a Sagebrush Festival in Crested Butte, instead of the Wildflower Festival."

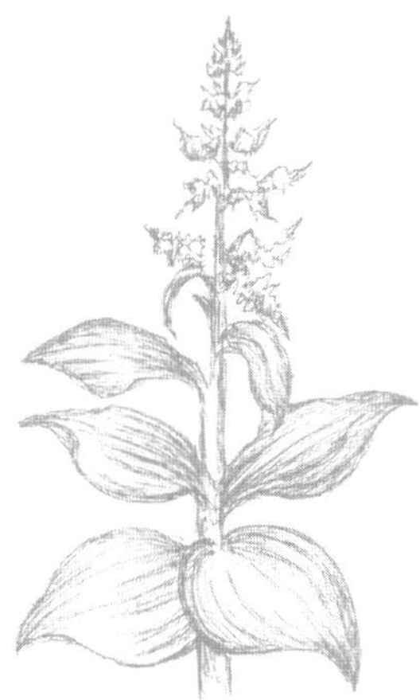

Mexican Cut Preserve - A "Research Natural Area"

Dr. Robert Willey — Dr. Scottie Willey

Not only is the Mexican Cut Preserve a unique and beautiful place, it is the center for a great deal of biological research conducted by investigators at the Rocky Mountain Biological Laboratory, and has been since at least the 1950s. It was acquired in 1966 by The Nature Conservancy – the first property in Colorado to be protected by The Conservancy – and leased to the Rocky Mountain Biological Laboratory, which continues to act as a "caretaker institution". In 1978, the Colorado Department of Natural Areas awarded it with its first ever designation as a "Research Natural Area". Additionally, the property was awarded "non-consumptive water rights" by the Colorado Water Conservation Board, something which is no longer permitted in the state. These rights designate the waters as being conserved for natural use, including scientific research, exempting them from demands for use by people. (Water rights in the American West are extremely complicated. Suffice to say, the water was safeguarded along with the Mexican Cut itself.) All told, 950 acres of land are protected at the Cut by The Nature Conservancy.[1]

As with most things at the Rocky Mountain Biological Laboratory, there is a story that leads to this outcome.

It was the year 1958. Two young biologists with Ph.D. degrees fresh from Harvard University arrived in Gothic as guests of Dr. Herbert Levi and his wife, Lorna. Dr. Levi was an invertebrate zoologist and, at the time, Curator of Arachnids at the Harvard Museum of Comparative Zoology. Levi had first visited RMBL in about 1947, when as a graduate student at the University of Wisconsin he was hired by Dr. John C. Johnson to help build cabins. While there, he took a course with Harriet Barclay and continued to return on an irregular basis throughout the 1950s and 1960s, building a cabin for himself and Lorna, which still bears his name.

Dr. Levi and Dr. Barclay had been doing research in the area of Galena Mountain, where the Mexican Cut Preserve is located. They encouraged the young couple – Robert and Scottie Willey – to visit the Cut, which they immediately recognized as special and embraced for their work. Bob Willey studied grasshoppers and Scottie's focus was dragonflies. RMBL and Gothic became their second home; they returned every year until 1996, when they retired and settled into life in Gunnison.

Meanwhile… in the mid-1960s Bob and Scottie began to get concerned about the private ownership of the land that was the Mexican Cut. They took it upon themselves to research and contact the owners of the patented mining claims in the area to see if they would be willing to sell. An inflated asking price and the advice of Dr. Levi led them to contact The Nature Conservancy, which did not have a permanent presence in Colorado at the time. Toward the end of 1964, The Conservancy agreed to put up money for the purchase of the land, through the auspices of RMBL. Working with the RMBL attorney, the Willeys brought the majority owner of the property to the table in August 1965 and an agreement on price was reached.

It took several more meetings and another year, but The Nature Conservancy agreed to assume ownership of the property and take responsibility for establishing tax-exempt status. In May of 1966, The Nature Conservancy purchased the first tract of land and designated the Rocky Mountain Biological Laboratory as the caretaker institution. RMBL also agreed to find donors who would be willing to repay the purchase cost to The Nature Conservancy. Repayment of these funds was completed in 1970. Additional lands were added to the parcel during the ensuing decades.

The legend that accompanies this business transaction is the fun part of the story – all the more so as it is true.

Scottie Willey to the rescue!!

After the signing of the initial agreement for The Nature Conservancy to purchase the land in August 1965, Bob and Scottie Willey were celebrating in their cabin in Gothic. Someone approached and asked if they were aware that a bulldozer was heading toward the Mexican Cut. Scottie flew out of her seat and behind the wheel of her pickup truck. As she approached the area leading to the Cut, she could not help but notice the marks left by the bulldozer. She travelled along the new "road" as quickly as she could and stopped her truck behind the bulldozer. Unable to get around it, she scrambled up a steep slope, through some trees and dropped down in front of it. Not exactly sure what to do next and uncertain if the driver saw her, she planted her ever-present "dip net" in the ground in front of the machine; she was too afraid to stand there herself. The driver stopped and asked her to approach his cab. She asked him what he was doing and he said he was clearing a road for exploratory mining operations. When she explained that the land was now owned by The Nature Conservancy and they had not authorized the building of a road, he very nicely gave her a ride to a turn-around place and left.

Scottie quite deservedly earned hero status for her quick response and actions. How many people would willingly stand down a moving bulldozer? The question remains, however – why was she carrying a dip net? In any event, Scottie saved the day. The road was later blocked to prevent vehicular traffic from passing, and remains blocked to this day. Access to the area is carefully marked, warning off random hikers and campers. The Mexican Cut remains as pristine an area as one can find in the Colorado Rockies, and is carefully tended by the investigators and students at RMBL.

Mysteries at the Mexican Cut

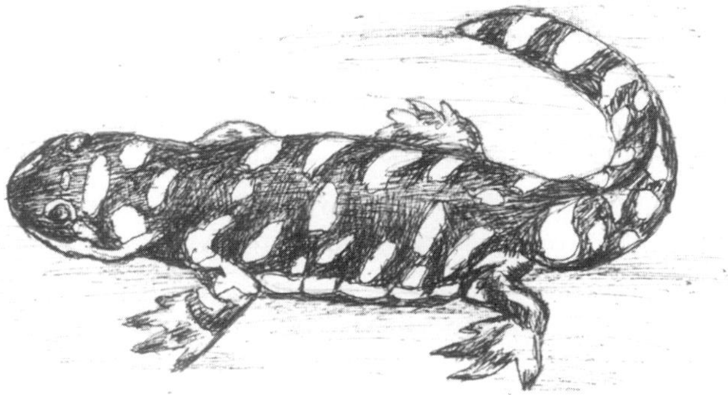

Dr. Scott Wissinger Dr. Howard Whiteman

LIFE exists on multiples of levels. People live, of course, as do dogs and cats and horses. So do trees and bushes and flowers, as well as butterflies and prairie dogs and salamanders and snakes and frogs. One must not forget insects and bacteria. All require the fulfillment of certain basic things to be alive, including nutrition and respiration. Reproduction is necessary for continuation of the species - all basic stuff that engenders curiosity for the biologist. How does a biologist decide to focus on life at the bottom of the food chain, maintain that focus for more than thirty years AND make it interesting to others? Ask Dr. Scott Wissinger.

When Dr. Wissinger completed his doctoral work at Purdue University in Indiana, he took a teaching position at Allegheny College, a small liberal arts institution in Pennsylvania. Although he had done interesting research in the American heartland while a graduate student, as an aquatic ecologist his curiosities were driving him in other directions. After consulting with colleagues at different universities, he chose to take the advice of Dr. Bobbi Peckarsky at Cornell and spend a summer in Gothic, Colorado at the Rocky Mountain Biological Laboratory. After only his second year teaching at Allegheny College, he was awarded a small starter grant and in 1988 he began his career-long affiliation with RMBL.

Water abounds in the lands around Gothic, so it seems a paradise for an aquatic ecologist – streams, rivers, lakes, ponds, waterfalls are everywhere. How does one choose? In Scott's case, he had been told by Peckarsky about the "wonderful ponds" at the Mexican Cut, so he was primed to visit them when he arrived. Little did he know where this journey would take him and the collaborations to which it would lead.

The Mexican Cut's special place in history has its roots prior to even the 1970s, when studies done at the ponds located there led to the discovery of the devastating effects of acid rain. These studies and the publicity surrounding the results (at least within the science community), as well as the focus and hard work of RMBL investigators such as Dr. John Harte, caused the Environmental Protection Agency to implement restrictions on emissions that would diminish, and hopefully do away with, acid rain. These restrictions have made a difference to organisms everywhere.

The Mexican Cut Preserve is located in a secluded area of the Colorado Rockies, not far from Gothic. Its pristine environment has made it the perfect venue for an array of biological research that has been conducted there for decades by scientists at RMBL. It was acquired by the Nature Conservancy in 1966 and has been continuously leased since then to RMBL to manage. At an altitude of more than 12,000 feet, the Mexican Cut contains a glacial cirque, which is a basin carved out by glaciers and which contains alpine ponds.

The ponds on one side of the Mexican Cut have unique characteristics and are home to a variety of aquatic species. Ponds on one side have some "buffering" from the soil, but those on the other side have none. Buffering occurs in ponds when the soil surrounding them contains a substance that absorbs and carries off acidic ions in the water, thus keeping them from becoming too acidic. Knowing this, Scott Wissinger thought that perhaps he would be able to detect differences between the buffered and unbuffered ponds using his research into aquatic invertebrates, hoping to add to John Harte's acid rain data. "I was going to study acid rain. I was going to be famous and show how acid rain affected aquatic invertebrates. That's why I came out here." Well, that didn't happen. But other things did.

Enter Wissinger's former undergraduate student and long-time collaborator, Howard Whiteman. Whiteman, an evolutionary biologist, was pursuing a Ph.D. and looking for a research topic. Wissinger, his former professor, suggested he might take a look at the salamanders at the Mexican Cut, which he did - and continued to do for decades, studying a unique life history trait of these interesting creatures. Intriguingly, although work on acid rain had pointed to the decreasing number of salamanders in the ponds, it became clear in 1988 that the population was rebounding. This did not make sense in light of previous research, so the challenge became to determine why.

There are really multiple science stories here. Each of these investigators has his own particular interest. Scott Wissinger, the aquatic ecologist, has focused on the aquatic insects that breed in the ponds and are food for the salamanders. Howard Whiteman, the evolutionary biologist, has been interested in the unique traits of the salamanders themselves. However, as one creature would not exist without the other, the two scientists have spent decades collaborating in their research. Partly this is due to the nature of the ponds themselves. These ponds manifest in three ways: permanent, semi-permanent and temporary. The species of animals that live in those ponds vary according to whether their homes are permanent, semi-permanent or temporary. That goes for both the salamanders and the aquatic insects – caddisflies in this discussion.

There is also the acid rain conundrum. This was one of the first issues addressed by Scott Wissinger when he began his work at RMBL, because of the unexpected rebounding salamander population. Funded by the National Science Foundation, he and Whiteman spent several years traveling to Gothic in early May. They measured what is called the "acid pulse" at the Mexican Cut, which happens every spring. What they discovered is that this pulse passes through the ponds each year before the salamanders become active. Acid rain certainly impacted the salamanders in the Adirondack Mountains and other places. So, what could possibly be causing the decrease or increase in the Mexican Cut population? Whatever the case may have been in the 1970s or earlier, by 1988 – and forward – the dynamics of the salamander population was shown to be tied to cycles in their own lives, and to natural peaks and troughs in their development cycles.

The life of a typical salamander begins as an egg and proceeds through various stages to full adulthood. It starts life in water and, assuming it completes its lifecycle, ends it on land. Whiteman began his research marking the salamander population in order to be able to follow it through its lifecycle. Initially, the results of this marking also showed that the population was growing on an annual basis. He was also able to observe that there were two different kinds of salamanders in the Mexican Cut ponds, one of which never left the water, which is very different from the "typical" salamander life cycle. The fully terrestrial, or "metamorphic adult" moved out of the ponds to land; the aquatic, or "paedomorphic adult" remained in the ponds and kept is larval morphology, including fleshy gils. This "dimorphism" is the result of environmental influences on the larva which determine whether the animal will transform into a metamorphic adult and move to land or remain in the pond as paedomorphs order to survive.

What is really interesting is that all of the salamander larvae have the potential within them to fully metamorphose. Food, predation and competition with other animals are factors that influence which ones make the move to land and which ones do not. Also, at the Mexican Cut the environment in a pond can be richer than that on land; more food is available and is easier to obtain. Therefore, some salamanders choose to remain as paedomorphs and enjoy feasting in the ponds while they are reproducing.

As food is an integral ingredient of this process, knowledge of the lifecycles of the aquatic insects on which the salamanders feed becomes vital. In steps Dr. Wissinger.

In the early phase of his investigations into the food webs at the Mexican Cut ponds, Wissinger noticed some patterns. In 'community ecology', species change according to certain 'gradients'. For example, species of flowers will change in a meadow based on the amount of moisture in a given part of the meadow. Dryer sections will host one species of yellow flower, while wetter sections will be home to another. This is also true in ponds, but what is it that determines why one set of ponds hosts one species of animal and another set of ponds, only meters away, houses a different species? This is what was observed with the aquatic insects at the Mexican Cut ponds. Not only did the salamanders differ in the ponds, but so did the food that nourished them.

The investigators began to notice that the ponds that were home to the paedomorphic salamanders, the permanent ponds, had a differing set of insect species than the temporary or semi-permanent ponds. They called this the "predator permanence gradient" – ponds that were deep and permanent and filled with predator salamanders contained one set of insect species; those that were shallow and temporary and dried up, contained another; and the semi-permanent ponds had a "mishmash of species". "More importantly," says Scott, although there was just one species of each insect that lived up there, each of those species had evolved certain survival traits". For example, the beetle genus A*gabus* in the permanent ponds was different from the beetle genus A*gabus* in the temporary ponds. "It was like evolution had fashioned a split where, these guys have traits that do well in permanent places and those guys have traits that do well in temporary places." They saw that for every kind of organism up there – little fly larvae, dragonflies, damselflies and caddisflies.

Caddisflies are similar to *Lepidoptera*, moths and butterflies, in that they build a case around themselves. "They go through molts just like caterpillars. They pupate. They seal up their case with silk, just like moths would, so they are protected while they rearrange their body tissues into adult tissues. Then, they fly out and look kind of like drab moths." Since caddisflies dominate the insect world at the ponds, Wissinger looked closely at the differences between those in the permanent ponds and those in the temporary ponds. How and why do they differ? And, if the temporary ponds dry up each summer, how can the caddiflies that live in them survive?

What he learned was that the caddisflies in the temporary ponds are "completely reckless" in their behavior. They live a fast paced lifestyle, driven to eat and grow quickly. They spend almost no energy on the silk they make to build their cases, which turn out to be thin. As the ponds begin to dry up, they become desperate to pupate and fly off. Thousands of them are pushed into smaller and smaller areas and they begin to nibble on each other's silk, which is rich in nutrients. The more a case is grazed on, the thinner it becomes, the easier for it be punctured by a nibbling neighbor. When that happens, the smell of blood in the water from the injured animal attracts mobs of other animals. They are cannibals; they devour their own kind. Since there are no salamanders in the temporary ponds, the caddisflies are their own predators. They themselves consume dead grasses which compost in the ponds over winters, but the silk from their cases is more nutritious than the grasses and the overcrowding in the ponds drives them to eat and grow, eat and grow.

What about the caddisflies in the permanent ponds? It turns out they are gifted with biblical patience. Not having to worry about their ponds drying up, they can take their sweet time growing. They build strong, dense cases, not easily perforated and too large for the salamanders to swallow. Because of the quality of their cases, they do not cannibalize and are safe from other possible insect predators, like beetles. They hide easily in the grasses from the salamanders and simply outlast their searching. Rather than pupate in July, they have the luxury of waiting until August.

That, however, is basically where the differences end. Both temporary-pond caddisflies and permanent-pond caddisflies eventually pupate. They conduct their mating ritual, with the females taking little packets of sperm called "spermatophores" from the males. The females then find a nice place in the trees to settle into a long "diapause", which is the term for a period of summer hibernation. "They shut down. It's like when R2D2 is turned off; all the lights blink and go out. They look dead." Sometime in early September, as day length decreases, the females wake up and lay their eggs under wood and rocks. Soon thereafter, the eggs are covered with snow and remain incubated at 32 or 33 degrees Fahrenheit all winter. Melting snow in the spring triggers embryological development and the lifecycle starts anew.

The males just die off or get eaten after mating. The females get to live into September.

What does this all mean for the future of salamanders and caddisflies? Well, they might continue to live in this happy (or not), symbiotic relationship forever. But, once again climate change rears its ugly head. The temporary ponds are now drying two to three weeks earlier in summers than when Wissinger and Whiteman began their investigations. As a result, the temporary-pond caddflies are forced into their frenzied cycle of eat and grow earlier and earlier. Because of their rush-to-survive-before-the-ponds-dry-up nature, they physically move at a much faster pace than the permanent-pond caddiflies. As such, they attract the attention of hungry salamanders, which find it much easier to ingest them than to hunt for elusive, slow-moving permanent-pond caddisflies. This does not bode well for the future of the temporary-pond caddisflies. As for the permanent-pond caddisflies, they continue to find refuge in their cases, which they use as little tents to hide them from the salamanders. Their future looks brighter.

The salamanders themselves continue to live their two existences. But what will happen to them once the temporary ponds cease to be around long enough to allow for their caddisflies to complete their lifecycles? If those easy morsels of food disappear and the salamanders are forced to look harder and wider for their nourishment, perhaps changes will occur within their population, too. And, oh yes, it must also be mentioned that cannibalism is part of salamander life in the Mexican Cut, too.

In the words of Dr. Wissinger: "We are starting to see species move upslope, so we are getting new combinations of species that have never been together before. Drying ponds are becoming so short-lived that the animals can't complete their lifecycles. The ones in the semi-permanent category are becoming more like the temporary ones. We are watching this to see if the species that live in temporary habitats replace those that live in permanent habitats. How will all this affect the system? For instance, the species in the permanent ponds processes more hay [their food source] than that in the temporary ponds. If you replace one with the other, the hay won't get processed and nutrients won't get recycled." Climate change is at the root of all this. How will the system be affected by climate change, which is already causing the species to change?

Questions continue to be generated and asked, but not always successfully answered. Time is necessary to tell.

On a happier note, however, the work done by Wissinger and Whiteman has had a positive effect on habitat mitigation in the United

States. Their discoveries about different sets of species living in permanent and temporary habitats have influenced lawmakers to consider that information when decreeing what kind of mitigation is necessary when a habitat is destroyed by development. If a highway is being built, or a developer is reclaiming land for a housing project, the powers-that-be now know to look at the nature of the land being developed. Is it wetland? Does it dry or not? What is its "hydro-period", the schedule by which it fills with water and then dries? Once this information is known, decisions can be made about what kind of mitigation is acceptable. To replace a wetland, another one must be created that fills and dries on the same schedule. Mitigation, which is replacement of the wetland, does not necessarily have to be done in kind, but knowledge of what is being lost helps to rebuild wetlands in another place. Were this not to be done (as in the past), the consequences to the environment can be devastating.

Just consider the Everglades…

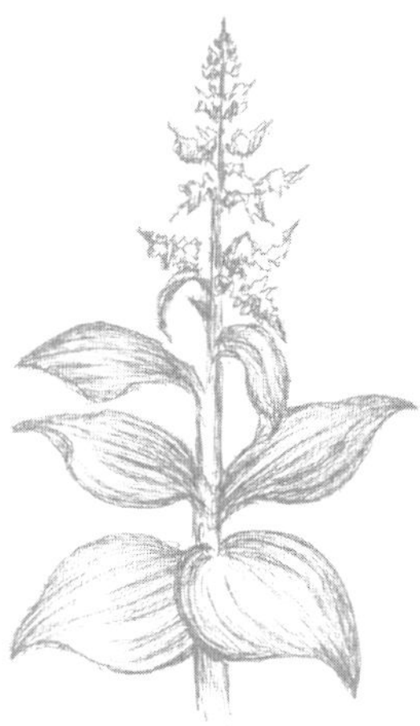

Hummingbirds and Hiking

Dr. William A. Calder, III (d. 2002)

Who doesn't love a hummingbird? These tiny creatures capture the imagination as they flit past windows at a gazillion miles an hour, hover just long enough at a feeder to lick up nectar to fuel them for the next excursion, chase each other away from dedicated food sources and occasionally treat a human to an eye-to-eye moment as they pause in front of you as if curious about this monster in their space. Then, they are gone. They arrive in Colorado and other summer haunts as far north as Alaska on schedule with the blooming wildflowers. Two or three months are spent moving from flower to flower, catching small insects on the wing (every animal needs *some* protein), nesting and reproducing before returning south for the winter. The birds and the flowers form a beautiful symbiotic relationship.

Hummingbirds and hiking have absolutely nothing in common except that in the mountains of Colorado hummingbirds proliferate in the summers, and hiking is a passion during which time one might spot hummingbirds. These two things also come together in understanding the person and the work of Dr. William A. "Bill" Calder.

[Dr. Calder passed away in April of 2002; the information in this chapter is garnered from friends, colleagues, written records and his wife, Lorene.]

Having received his PhD from Duke University in 1966, the young Dr. Calder worked in several positions before settling in at the University of Arizona in 1969. For a time, summers were spent doing research near Jackson Hole, Wyoming where the University of Wyoming has a field station. In 1969, however, a colleague who worked at the Rocky Mountain Biological Laboratory in Gothic, Colorado invited Bill and his wife, Lorene, for a short visit. They became hooked by the possibility of research on hummingbirds, which had become a new passion, and by the ability to teach classes. The Calder family returned to Gothic in 1971 for their first full summer. The historic 'Richards' cabin they lived in then and for many summers was quite old and basic, but they had it to themselves. A wood stove to heat and cook on completed their roof and walls; Bill would haul dead branches home from his daily run and saw them up for firewood. The two children were free to run amok with the children of other resident scientists. Lorene could – and did - work alongside Bill in addition to pursuing her own interests. It was easy to settle into life in Gothic.

Bill Calder was a physiological ecologist. In lay terms, that essentially means he was interested in how animals in their natural setting survive and thrive. What do they eat; where do they nest; how far afield do they go for food; how do they keep warm during cold nights or cool themselves during hot days? The natural setting, the great outdoors, is important in determining answers to these questions, not how the birds can be observed or manipulated in a laboratory. Why hummingbirds? In Calder's field of study it is important to focus on extremes in order to discover what the "physiological machine" is capable of to foster the survival of individuals. For example, Bill was interested in the thermal ecology of hummingbirds; essentially, how do they keep themselves warm enough to live in an environment that turns cold? Gothic and the surrounding Colorado mountains are known to experience quite chilly nights during summer months. This was probably more true in the 70s and 80s, when Bill first began his research there and climate change had not yet made a real impact. Through his research, it was discovered that hummingbirds actually have the ability to lower their body temperature overnight to a few degrees above the outside temperature, and – even more exciting – to bring that temperature back up in the morning, exiting the overnight torpor at precisely the time they would have naturally had the night remained warm. Quite an achievement for a tiny little bird!

The species of hummingbird that makes Gothic its annual summer home and was Bill Calder's research focus is the broadtail. Bill was interested in many aspects of hummingbird life, including nesting patterns and weight gain and migratory habits. In order to follow these tiny creatures accurately, it was necessary to capture and band them. Bird banding is closely regulated by the government, but Bill was a licensed bander. Imagine how small and lightweight a leg band must be in order to fit and not hinder the comings and goings of a bird that size. The birds must be able to maintain their balance when flying and perching, and nest without disruption. But the banding was essential to the work, so band he did. Tales are told by those who still remember Bill enticing little birds into traps he built by his house. He would capture them, weigh and measure them, band them – and return them to Nature. In order to facilitate his work, several other then-graduate students became licensed to band, including Nick Waser and David Inouye, banding as many birds as they were able. The fun part of the research was seeing the same birds returning the next year, or the next, or the next carrying those bands. As a result of this work, much is now known about "site fidelity" – that the birds migrate thousands of miles each year and return to the same place each summer. Interestingly enough, however, Calder was never able to determine specifically where "his" birds migrated FROM each year. Although he followed many birds in Mexico, and broadtails in particular, none were the ones he banded in Gothic.

It is understandable that if one is studying extremes in Nature, the size of an animal would come into play. Bill Calder became interested in size effect, or scaling. Say that you look at a mouse next to an elephant; what changes in the physiology and morphology of each animal to accommodate the needs of their extreme sizes, be they small or large? In birds, one could place a hummingbird against a large bird like a kiwi, something that Calder, in fact, did. He actually traveled to New Zealand to study kiwis because, relative to their size, they lay the largest egg of any bird. And, they lay only one egg at a time, which is almost as big as the poor mother!

Nesting physiology was another area of study for Calder. When one hears the distinctive "hum" of the broadtailed hummingbird, it is the male making the noise with a modified feather in each wing. Female hummingbirds are much quieter, although they do emit small chirps as they go about their business of finding food or seeking the exact branch under the exact tree canopy they need for a successful nest. Combining that silence with their size makes it quite difficult to track the females. Interns, such as Nick Waser in 1972, and other helpers spent many a frustrating hour listening for female chirps in the forest in search of elusive nests. When they did find them, they were charted as to location and the number of eggs was counted. But this was difficult work. So, Bill became interested in trying to find a way to monitor female nesting. Fast forward to forty years later…

One of the most interesting aspects of life in Gothic is that the scientists, all specialists in their fields, become generalists as a result of living – eating, drinking, hiking, backpacking – amongst each other. They tell stories around the campfire or use each other as sounding boards. The result is that a task which was the brainchild of one frequently becomes a collaborative project for the many. Such was the case with Bill Calder's hummingbirds. What would a hummingbird be without flowers and their nectar (and those little insects for protein, of course)? How would a flower survive without its pollinators?

As it was difficult to find a reliable way to track the females and their nesting habits, the question was essentially put on the back burner for a number of years and the labor intensive work of people climbing trees and counting nests continued – and continues. But the world has changed a great deal in forty years. Technology abounds! It came to the mind of David Inouye, a specialist in wildflowers and a colleague of Calder's at Gothic for decades, that technology might be used to track the female hummingbirds. What about attaching a micro radio transmitter to the backs of the birds? That would work, wouldn't it? So in 2014, Inouye set about experimenting first with some males, concerned that anything attached to a female might interfere with nesting. It seems, though, that the male hummingbirds disappear at night. Where do they go? Several were captured and tiny radio transmitters were attached to their backs. It was learned that they tend to go up, into the higher country, to sleep. However, keeping the transmitters attached to the birds turned out to be quite a challenge. They can't be clamped on – the clamp added to the weight of the transmitter would be too heavy for the bird to bear. So they must be glued on using, it turns out, eyelash glue. But glue can fail. The transmitters are quite expensive and many an otherwise productive hour must be used to track them if they fall off. It seems that this particular project has merit, but will require time, diligence, creativity and money. Naturally.

Fortunately, hummingbirds are not as frail as they appear. They can't possibly be or how would they manage to fly thousands of miles, twice a year, just to find some nectar? Butterflies migrate those distances, too, but over the course of generations. A question that begs answering is: How long do hummingbirds live?

The banding done at Gothic shows that hummingbirds can live for many years. The record longevity for a broadtail hummingbird is minimally twelve years! It was a female, trapped and banded in Gothic in 1976 by Nick Waser. She was an adult, possibly just a year old but possibly already older. The number on her tiny aluminum band was X18025 - not a very beautiful "name" for a pretty little bird, but that is what she continues to be known as. It is quite possible that she lived longer than this, but the last time she was captured and recorded was eleven years after her banding. "That", says Nick "is off the charts for a bird that weighs three and half grams! Way off the charts!" Hummingbirds are way above average for lifespan in birds. "They live way too long for how big they are." Of course, not all individual broadtails (or other hummingbirds) live that long – but we are talking about extremes.

Owners of hummingbird feeders are aware that commercial nectar sold for them is dyed red. It is often assumed that hummingbirds are attracted to red. That assumption is erroneous. Tests done dating back to the 1940s prove that hummingbirds have no innate preference for color. This does not mean they are colorblind, just that they are able to rapidly and correctly discern where the nectar is abundant in spite of the color of a flower. In Gothic, the favorite wildflowers of the broadtails include low larkspur and scarlet gilia, also known as fairy trumpet. Larkspur is purple-ish; scarlet gilia is red-ish. The birds arrive for the blooming of the first and remain through the blooming of the second. Well-meaning humans enthusiastically provide feeders filled with sugar water, which compete with the flowers, but the birds really prefer their nectar straight from the flower. If natural nectar is healthier for them than sugar water, it has yet to be proven, but in years of excellent bloom the birds have been scarcer at the feeders and thus more difficult to trap.

Bill Calder was a conservationist and as such was interested not only in the local aspects of hummingbirds, their behavior and survival, but also about their regional movements. He trapped locally not only in Gothic, but at Emerald Lake, Washington Gulch and other surrounding areas. He traveled to Alaska to study the rufous hummingbirds that migrate to and nest there. He also spent a good deal of time in Mexico, working with local scientists there, helping to raise awareness about the birds and learn more about their migrations. Lorene notes, "The expertise and experiences Bill shared so freely with his Mexican colleagues helped to establish a very strong hummingbird monitoring program in the Mexican states of Jalisco and Colima. One of our young Mexican students of high school age was a natural bander of hummingbirds and has become a prominent Professor at the University of Guadalajara, specializing in [hummingbird] ranges and population dynamics," and collaborates closely with hummingbird researchers in the United States.

Lest we forget the hiking part of this essay… Bill Calder was an avid outdoorsman. He was known to have ridden his bicycle from his home in Tucson, Arizona to Gothic. He taught many a field class along the hiking trails surrounding Gothic. Whether the subject matter was ornithology or something else, his classes in general came to known as "Backpacking 101". In this way, he influenced many young students to experience the mountains and fields and streams and lakes of the Colorado high country.

A remembrance from 1972 places Calder at the head of a group of graduate students, newly arrived and anxious to begin their work. Instead, he insisted they "go for a hike". It was June and there was snow on the ground in many places. The agreeable group, Lorene included, followed Bill up the long path to Triangle Pass (also known as Conundrum Pass), slogging through snow for the last mile or so. They arrived at the pass, altitude about 13,000 feet, late in the afternoon and Bill suggested they should all go to the hot springs for a dip. Exhausted and more realistic than their mentor, the students prevailed upon him to return to Gothic. This illustrates his bountiful energy and the enthusiasm he exhibited for his surroundings.

In a memorial written after Bill's death in 2002, his friends and colleagues Nick Waser and Mary Price wrote, "At the Lab, Bill carried out research on hummingbird physiology and energetics that brought international recognition, always taking time in between to hike the highest peaks and passes… he [was a successful] mentor of young scientists at all stages of their careers and from any corner of the planet he strove so hard to conserve… Bill Calder provided us with a powerful model for how to live and how to do science: equal parts of humor, keen observation, hard work and absolute honesty."

And the hummingbirds continue to come to Gothic.

"NickandMary"
The Birds, the Bees - and Breaking Tradition

Dr. Nick Waser Dr. Mary Price

We have all heard tell of people who are so connected they seem to read each other's thoughts and tend to finish each other's sentences. An idea begins in the mind of one, translates into words tumbling in bits out of his or her mouth, only to be seized by the other and the words turned into completed phrases, the idea crystalized into focus. Thusly we meet two long-time scientists at the Rocky Mountain Biological Laboratory who, in many ways, define and are defined by the place, Dr. Nick Waser and Dr. Mary Price - known to the community as "NickandMary".

Nick Waser first visited Gothic in the summer of 1972. He had learned of the Rocky Mountain Biological Laboratory through a college friend at Stanford and, being in that place in life where one must choose a path, he decided that some time at Gothic might be a way to broaden his horizons. As did many a new student at RMBL, he wielded a hammer to help repair buildings, in addition to interning with a researcher and doing some of the research leg work. In Nick's case, he climbed trees to seek out hummingbird nests for Dr. Bill Calder, from the University of Arizona, a task that served to develop his interest in hummingbirds. This experience also helped him to realize that "hummingbirds are hard to work with; it is easier to work with the flowers that they visit and pollinate". Thus began his life's work with pollination and pollinators. He decided to attend the University of Arizona to pursue his Ph.D., and there he met Mary Price.

Mary Price began her graduate studies at the University of Arizona in 1971, a year before Nick showed up. Her field of study was desert rodents, but her interests were diverse. She and Nick "connected" almost immediately upon meeting and their professional lives became intertwined along with their personal life. In early 1973, Price attended a field course run by the Organization for Tropical Studies (OTS), which had been founded partially through the influence of RMBL alumna Jean Langenheim. The teaching assistants in the course were people associated with RMBL – David Inouye and Graham Pyke. So Mary had had an introduction to RMBL through her connections in this course. During the summer of 1973, she went to Gothic to visit the people she had met through OTS. It was to be just the beginning of a lifelong connection and collaboration with the people who are RMBL.

Finding their way via interesting post docs through a myriad of fascinating biological species – hummingbirds, desert rodents, *foraminifera* (protozoans), low larkspur, scarlet gilia (fairy trumpet) – Nick and Mary consistently scheduled time to spend in Gothic during summers. Eventually, their dedication to working together as an academic couple led them to the University of California at Riverside and a position that they "sort of shared" for twenty-five years – two scientists, one lab, two specialties. They made it work and, in many ways, are models for the modern concept of job-sharing.

Back to Gothic… Before discussing the work they have focused on over the years, an aside about the culture in Gothic and their role in changing it is necessary. The Rocky Mountain Biological Laboratory that Nick and Mary and everyone else arrived at in the early 1970s was, essentially, still a family-run institution. There were standards of conduct set by the Johnson family, quite conservative in the changing environment of the day. Nick and Mary were already a couple in Arizona and did not wish to spend their time in Gothic having to live apart, so Nick applied to the Board of Trustees to be allowed to cohabit with Mary. The Board agreed, but at the same time did not desire to oppose the wishes of the Johnsons. A solution was offered by RMBL's attorney; have Nick and Mary sign a piece of paper (which they still cherish), drawn up by the attorney, which said they had "the habit and repute of being married". This did satisfy some of the conservative members of the community (although not all). The couple was assigned "a tiny cabin and the prediction was that we wouldn't last the summer as a couple. The cabin was just so little!" The object of forcing them into a cramped space was to prove that they "only wanted to shack up" and wouldn't survive the summer in such close quarters. Little did anyone know!

The reality of the situation was that the document Nick and Mary signed actually did marry them. Marriage in Colorado then, as now, required only a document signed by both parties (and today it must be registered). Not knowing this at the time, Nick and Mary did get married again a year later in Arizona. However, this story is legend in the Gothic community. From this point forward, more and more researchers chose to house with their partners, Bobbi Peckarsky being one. She and her future husband asked Chris Johnson if would agree to them sharing a cabin. He disapproved, but finally bowed to changing mores.

Onward to the research!

It is not uncommon to hear Nick and Mary describe themselves as "generalists". The focus of Nick's research has been the relationship between pollinators and the plants they pollinate. He began with hummingbirds and expanded his work to include bumblebees and other insect pollinators. Mary was and is a mammologist, studying desert rodents such as kangaroo rats for many years. When discussing their work, however, one is immediately drawn away from the specifics of pollination or mammal behavior into a world of the interconnectedness of Nature.

Nick strongly believes in the importance of understanding pollination and the pollinators that effect it. To be succinct, "The study of pollination is important because we eat the product, and because the health of natural ecosystems depends on pollination by animals. Ninety percent of the plants of the world make use of animals, as opposed to wind, for their pollination. So it is very important to understand this interaction between animals and plants." The work done by Nick and others, including his students and post docs, have helped to make the Rocky Mountain Biological Laboratory an international center for pollination work.

Mary chose to focus her personal research for years on a question about the co-existence of desert rodents because, there is "a great diversity in things that are ecologically very similar to one another", and she was interested in why one species doesn't outcompete all the others. A post doc offered her a change of direction for a while, looking at *foraminifera*, "of all things", in the salt marshes of Nova Scotia. *Foraminifera* are "little protozoans that make the White Cliffs of Dover".[1] Always interested in the distribution of species, Mary studied the zones where these creatures resided. Once settled at UC Riverside, she returned to working with desert rodents. She and Nick traveled to the southwestern US deserts – the Mojave, the Sonoran, the Chihuahuan, the Great Basin – to collect animals for study back in the lab, where they developed "quite a menagerie". Realizing eventually how hard this was on the animals – once they are separated from their territories they cannot be released back to nature – caused Mary to address the work more in the field than in the laboratory.

So, assuming that Nick Waser is a "specialist" in pollination and Mary Price is a "specialist" in desert rodent behavior, what exactly to they mean when they call themselves "generalists"? Nick's take on a specialist is an individual who is absolutely THE go-to person on any given subject. That is all, or mostly all, that person works on. Using Mary as an example, however, he points to the fact that she has worked with a variety of different mammals; but, she has also been interested in plants. "She is a very good field botanist, but when she came up here she began to work with hummingbirds and bumblebees and pollination of flowers. She's thought about everything from ecology to physiology to morphology to animal behavior. THAT's what I mean by a generalist." She also continued her work with rodents, such as the deer mice that plague the Gothic area.

Nick continues with his "generalist" definition. "If you want to go somewhere, Nature leads you with the questions She presents you from one area, to the next area, to the next area. You often find the really stimulating ideas where the different fields of thought, the traditional fields of thought, come together – where ecology and evolution come together, or ecology and animal behavior, or ecology and physiology come together. You may have to collaborate with somebody in a different field, because you are just not *specialist* enough to get at the depth of what one does. That is why you form networks of collaboration. That's what we mean by *generalist*. The generalists may collaborate with the specialists in the different areas and that may be very, very profitable for everybody."

Having said all this, what has been the focus of Nick and Mary's work at RMBL?

Nick's research has evolved from his early dissertation topic on the interaction between flowers and birds, plants and pollinators. Early on, hummingbirds were the pollinators he chose to look at, partly because of the fact he had become familiar with them as a graduate student working with Bill Calder. Hummingbirds, though fascinating creatures, are difficult to pin down and track in the environment. But the flowers they pollinate return year after year in the same places. So Nick became interested in the two major food sources for hummingbirds, the low larkspur and the scarlet gilia. These two plant species tend to flower in sequence, but some years they overlap in the timing of their flowering for a week or two; other years, they do not overlap at all. What happens during years of overlap? Do the plants suffer? It turns out that they do. Hummingbirds tend to be pretty indiscriminate in their choice of natural nectar, and they move freely from one plant species to another that is in bloom, mixing up the pollen. If a plant doesn't get as much of own species of pollen, it does not make as many seeds. Some years later, the question of whether a reduction in seed production has an effect on the next generation of plants became a topic of study at RMBL for Nick and Mary, along with some of their colleagues. They spent about fifteen years exploring the question, and it seems that the answer to their question is… yes and no!

Together with Mary, Nick decided to expand his study of pollinators to bumblebees. The low larkspur/scarlet gilia project had identified hummingbirds as the pollinators. They decided to look at bumblebees as pollinators of larkspur, and realized that they also pollinate scarlet gilia. Along the way, they became quite enamored of bumblebees!

Not turning her back on mammals, Mary had attended a seminar by Otto Solbrig of Harvard University when she was almost finished with her dissertation. In the seminar, Solbrig spoke about what is called the "patchiness" of the environment and how plants adapt to that. She and Nick looked around Gothic and decided the environment was a patchy one – it differed from spot to spot. What, they wondered, if there was local genetic differentiation? What if the seeds from a plant are pollinated by pollen from a different environment? Will the seeds be well adapted to their home environment? Heady stuff!

The result of all this thinking about pollen and where it comes from and how it is delivered was that Mary spent that particular summer doing one of the first hand-pollination studies at Gothic, using pollen gathered from plants growing different distances away from the focal plants. This started her and Nick on a fifteen year study of the issue of local adaptation. Along the way, they became interested in other things such as whether it matters how many seeds are produced by a plant, the color of larkspur flowers and other mysteries of Nature.

Forty plus years of working in one environment allows for a certain amount of devotion, not only to the place, but to the work that has been done there. The building of data sets and making them available for future researchers in the same place help to define what has become known as "the ecology of place". A book by that name was published in 2010, edited by Mary Price and the Director of RMBL, Ian Billick.[2] Mary says of the concept, "it is not a surprise that the ideas about ecology of place emerged out of this institution. People here have a sense of building on historic data sets, building on what people have seen before and adding to it in an interdisciplinary way. Some characteristics of this Lab have facilitated an intellectually profitable way of studying ecology." Nick adds, "To some extent, the stories in this book illustrate how complicated science is. It doesn't fall into the neat categories philosophers of science talk about. You do go out into Nature with some expectation of what you'll find. But, if you don't allow Nature to tell you, sorry, that's not the way it's working here, and here is today's surprise, then you don't make much progress. If you do science by following your nose, to some extent, and by following serendipitous events, in the end you reach a level of understanding of how things work."

And follow their collective noses they did. Continuing with the theory, Nick says, "The nice thing about coming back here is that it really does take a long time to follow your nose on these projects, and to accumulate enough experimental evidence to be convinced that you've got a pattern." To follow one's nose is to be diligent and to pay attention, and to take one's time. Starting with flowers, one looks at pollinators – bumblebees and hummingbirds. Questions can be asked about how the bees or birds decide which flowers to visit. Those questions are about animal behavior, so from there more questions can be formulated about what goes into the decision. Is it the color; the scent; the size? Experiments can be designed to manipulate those things, all the while allowing the animal to lead you in the direction of the answers.

They followed their noses in this way not just on one project, but on a series of projects dealing with the behavior of animals and their attraction to specific flowers as constituting natural selection, or part of natural selection as it relates to the characteristics of the flowers on the plants.

From there, broader questions can be asked about the distribution of flowers of different kinds, about different species within a meadow. What is the genetic composition of the meadow? A meadow filled with many species of flowering plants generates questions about the composition of the whole community of plants, as well as the animals. Questions such as these led them and others at RMBL, to look at pollination of the meadows as a network of interactions, a web. They've learned that it is a web, something not known a decade and more ago. Such work has been picked up and followed up on by scientists in other parts of the world, leading to collaboration with other researchers. What an exciting turn of events!

But just a minute! Are bumblebees and hummingbirds the only pollinators; or the only ones worth studying? No. There are also, flies, moths, butterflies and beetles. Anything that goes from flower to flower, looking for something to eat, moves pollen around - and helps to ensure the reproduction of the plant. These, animals, too, are brought into the scope of the studies. It also becomes critical to understand the importance of the role of insects in the life of plants. Killing off bees or pollinating flies or butterflies is not just a bad idea – it is counterproductive to the health of the ecosystem.

One result of years of nose-following has been a growth of understanding and love of insects such as bees and pollinating flies. It is easy to love a butterfly, because butterflies are beautiful and magical. Bees sting. Flies bite. Some do, some don't. Nick and Mary have developed a deep esteem for these little creatures, exclaiming over their beautiful wings, or coloring: "They are gorgeous! They look like little hovering jewels!"

The extent of the interests of these two long time RMBL researchers goes on and on and on. Habitat selection. Effects of deer eating wildflowers. Seed versus plant consumption by animals. Ecological risks of genetic modification. Children as natural scientists. The growing interest in genetics. The importance of all scientists to document their work, share it with their communities, safeguard it for the future.

Talking with Nick and Mary is like entering a universe of dialog about the universe. The Rocky Mountain Biological Laboratory and Gothic, Colorado are a big part of their universe, but it continues to expand.

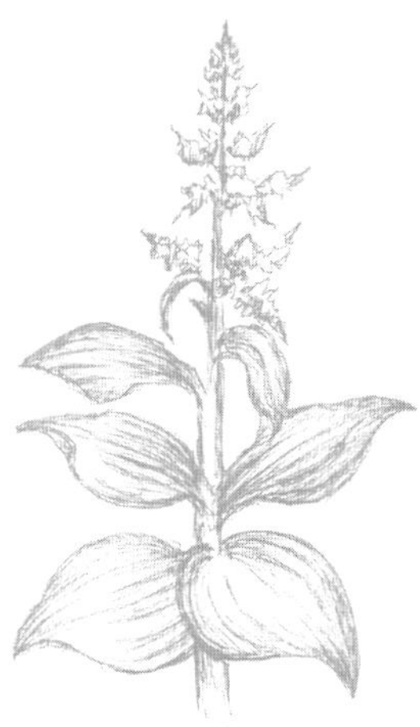

G'Day Mates
RMBL and the Australian Connection

Dr. Andrew Beattie Dr. Graham Pyke Dr. Christine Turnbull

Strolling through the Rocky Mountain Biological Laboratory and the little town of Gothic one might understandably conclude that being tucked away high in the mountains of Colorado would preclude international interaction. Who, in a pre-internet world, would have known much about the place, the work done there or the people who populated it? Why would it have been important enough to have caught the attention of someone in Europe, Asia or Australia? What would bring scientists from those places to this small town located at a challenging altitude? In the twenty-first century, it is difficult, if not impossible to hide from the world, but it wasn't always that way. And RMBL was, for the most part, hidden. Word of mouth was its greatest advertising medium. This little gem, unaffiliated with any university or government institution, was kept alive by networking scientists who shared it with those whom they thought would benefit from its location.

The late 1960s and early 1970s were a prime time for the introduction of new biologists to RMBL. For the most part, the troubles of the "Gothic Wars" and transitional years were over and a new generation of young scientists was being introduced to Gothic by those who had stayed. Two young British scientists and an Australian were among the diverse researchers who arrived each summer.

Andrew Beattie was trained in the United Kingdom and came to the United States in 1968 to work at Stanford University. His first summer there he joined with Paul Ehrlich and his group of investigators to spend the summer doing field work in Gothic. "That," he says, "is a story in itself. We arrived on the first of July, which is when the butterflies are usually flying, but we had three days of snow. No butterflies appeared that year, so we all had to do other things." This initial introduction to RMBL was enough to entice Andy to return for many years, although there were long gaps between his visits. During his years in Gothic, however, he was involved enough to serve as the last part-time Director of RMBL, and responsible for convincing the Board of Directors that it was time to hire someone full time.

Christine Turnbull, also educated in the UK, came to Gothic with Andy a decade or so after his introduction to RMBL. Christine had completed her undergraduate work at the University of London and moved on to earn her PhD at Northwestern University in Chicago. Both Andy and Christine were and are interested in social insects, such as ants and bees, and the early work they did at Gothic had to do with how ants disperse seeds. Their findings showed that ants provide dispersal services for a large number of plants. In 1987, the couple moved to Australia to work at Macquarie University in Sydney, but continued to return to Gothic every few years to do research and to allow their children the experience of growing up with the unique freedoms that the area allowed.

The couple's commitment to the success of RMBL was manifested in many ways. During her early years in Gothic, Christine served as graduate student representative on the RMBL Board of Trustees. In that capacity, she worked with "a handful of RMBL residents" to develop and stimulate what would become a significant fundraising effort for the laboratory – the merchandising and retailing of RMBL-related products. Some of her tee shirt and sweatshirt designs sold out year after year; some continue to be available in the RMBL Visitors Center gift shop. This area of RMBL activity began to supply a welcome stream of revenue during her tenure, and has continued to do so in years since.

Beattie remembers well his efforts to encourage the transition from part-time to full-time Director – "based, I should add, on the bitter experience of three frantic years being both Professor and Director! I remember the Board meeting when I suggested this, as it represented a very big change in the administration of RMBL".

Andy was the motivator behind RMBL's initial applications to the

National Science Foundation for facilities, as opposed to research projects, and was Director when a serious effort was made to raise funds from business and industry. "I actually had to buy a suit (!) to tread the deeply carpeted floors of corporation head offices and government agencies, to persuade executives (with much more expensive suits) to give to RMBL. On one occasion I visited the CEO of a very well-known soft drinks manufacturer and he asked, immediately after introductions were completed, how much I wanted. As I pondered this he pulled his checkbook from his inside pocket and looked up expectantly. This was a moment of truth and I named a figure. He wrote the check immediately. I cursed myself as I had clearly aimed too low!" He acknowledges that the "peak event was when the Governor of Colorado, Dick Lamm, hosted a major fund-raising event for RMBL at his mansion in Denver".

Andy and Christine focused on building good rapport with the local Crested Butte community, especially the cattlemen. A result of these efforts was annual gifts of homemade sausages, especially venison, which they happily donated to the RMBL kitchen for all to enjoy. But, the main goals during Andy's tenure as Director were "to develop RMBL as an international science center and to improve its teaching and research facilities". To this end, he initiated the Scientist in Residence program, which brought in distinguished ecologists from around the world to teach senior undergraduates and graduate students for two to three weeks, believing that "one of the greatest benefits of these occasions was the opportunity for students to meet great scientists in informal settings, including fieldwork and evening barbecues." He also arranged for RMBL to host a gathering of the Society for the Study of Evolution in Mt. Crested Butte, one of the first major scientific meetings to be hosted by RMBL.

In addition to his service as Director and his work as an investigator, Dr. Beattie took an interest in the practical needs of RMBL. The cabins were old and the danger of fire was a persistent concern. One day, when speaking with people from the local fire department, he was told that there was an old engine available for sale. "Do you want it? I said, sure. How much? $250. I said, does it go? It went. Its tires were intact. So we drove it up here. I bought it for the Lab. It was stationed here. We got everything working, got the pumps going. We squirted water over people and over cabins. It had a big tank on it, a hose and everything!" The fire engine was maintained by the work crew but it became too large a burden as the winters took their toll, and it became increasingly difficult to revive the engine as the years went by. As there were no fires, the truck was not utilized for that purpose, but its presence was comforting to the community at a time when local firefighting resources were scarce.

Since 1968, Andy and Christine have continued their professional and personal connection with RMBL investigators, including Paul Ehrlich, with whom they published a book in 2001 titled ***Wild Solutions; How Biodiversity is Money in the Bank***.[1] Based on years of work together and apart, at Gothic and around the world, the book discusses the millions of species that provide ecosystem services and maintain the quality of our air and water, as well as the fertility of the soil. They show how the creatures dispose of domestic, industrial, and agricultural waste, and protect crops from pests. Also examined is how biological diversity opens the way for new medicines, pharmaceuticals, construction materials and designs, and manufactured goods. Their latest research has focused on antibiotic secretions by ants and other social insects that may have major importance to humans.

Not wishing to "bignote" themselves (as they say in Australia), Andy and Christine believe that working to create a full-time administration and to reach out further into the international scientific - and business - communities helped establish RMBL firmly on the world map.

Graham Pyke arrived in the United States from his native Australia in 1970. His first time ever abroad, Graham had enrolled at the University of Chicago to pursue his doctoral studies. An ornithologist at heart, he traveled to Costa Rica in early 1972 to attend a course offered by the Organization for Tropical Studies.[2] For two months he "lived, ate and breathed tropical ecology" and discovered he was fascinated by bees. While there, he met David Inouye, who convinced him to visit Gothic, because there were "these flowers and meadows and lots of bees and the bees are easy to watch and it's a great place". Coincidentally, he learned that one of his post-doctoral colleagues in Chicago, David Allan, was planning to spend the summer in Gothic doing research on what bees eat. They decided to travel together and arrived at RMBL in June of 1972.

"I discovered immediately that things were not quite as I had imagined. David had created a picture of basically a two-dimensional world, with flowers and bees; but, actually, it's a three-dimensional world. There are plants, and the plants have structure, and the flowers are in a three dimensional arrangement. I started to do research on that." Graham returned the following summer, finalizing his doctoral dissertation and, by the end of that time he had "kind of fallen in love with the place".

Accepting a position at the University of Utah in 1974, the new Dr. Pyke negotiated some start-up money for research and the use of the university van. Utilizing these resources, he returned to RMBL that summer and began a survey of bumble bees and the flowers they visited, covering the area from the center of Crested Butte to the tops of three mountains in the valley. The survey lasted for a period of three months and each site was visited weekly – "we did ten thousand vertical a week, so we were very fit!"

The time was ripe for research grants and Graham was fortunate to be able to return to RMBL for thirteen consecutive years. Although he chose to return to Australia in 1977 and to build his life and career there, he was able to continue his research in Gothic until 1985. All told, he has spent 18 years visiting and doing research at the Rocky Mountain Biological Laboratory – with a twenty-two year hiatus between 1985 and 2007.

Graham first met Paul Ehrlich in 1972 but it took two decades for the two to become fast friends and professional collaborators. Pyke's focus on bees and Ehrlich's passion for butterflies did not deter them from capitalizing on a shared love for birds. Annual birding trips have blossomed into a joint project about nectar-eating birds in Australia and Papua New Guinea. They have also worked together on the importance of maintaining biological collections to address ecological environmental issues.

"Two hundred years ago, whereas the general community said to go off to the far corner of the world, like Australia, and bring back some amazing things, like a platypus, and we will be happy – that's gone. People are no longer willing to support that kind of stuff. Back then, discovery and amazement were important. Now people are really concerned about pollution, loss of biodiversity, climate change, the fact that we humans are doing a great job of stuffing up the planet. Biological collections need to be in that direction, if we are going to survive." Money to maintain the huge collections in the world's great museums has been drying up, so it is growing ever more important for the scientists themselves to make certain their work is archived and collections are maintained.

Archiving data at RMBL is what Graham Pyke calls "something that gives me an excuse to come to this part of the world". Since 1928, a "huge" amount of research has been done. Unfortunately, much of it "resides in people's notebooks, or has been tossed out". Working in collaboration with RMBL Director Ian Billick, Graham has been attempting to obtain data from other scientists and turn it over to Ian to be archived. "The [Bob and Scottie] Willey data was my big success. They were very keen on passing their data over to me, and I passed it over to the Lab." This can, however, be quite difficult to achieve. "If people are still alive and doing research, they are just so busy they can't take time out to lift their data from the box it is in on the shelf, put it in the car and drive it to RMBL. I am talking about raw, scientific data. Archiving is a really important issue because once people die their data often dies with them." Graham Pyke has archived his data, as has David Inouye and a few others. He continues to encourage people to do so. Unfortunately, much has already been lost.

In 2005 and 2006, Graham was contacted by RMBL scientists David Inouye and James Thomson. They were interested in having him return to Gothic and repeat the surveys he had done in 1974. So, after twenty-two years, Pyke did just that, working with his colleagues. This time, however, the surveys were conducted in the context of climate change. Using data that showed there had been changes to the climate over the course of that thirty-four year period, they made the assumption there would have been "associated changes in the distributions of bumble bees and possibly the plants they visit. We expected to find – and we did find this to a degree – that bumble bees and plants have moved higher up the mountain." These findings resulted in Graham getting "sucked into climate change", and encouraged him to teach a course at RMBL the following summer with John Harte and Lee Panetta on the topic. He has also partnered with Paul Ehrlich and others in the *Millennium Alliance for Humanity and Biosphere* (*MAHB*).[3]

First designated in 2004 as a "highly cited author", Graham Pyke encourages scientists to write well and often and to use citations when doing so. When asked what it means to be a scientist, he offers two main sets of ingredients. "The fundamental ones are a kind of curiosity, a desire to understand the way the world works out there. But the second one is a desire to tell the world what you find. There are lots of people who are inclined in one direction or the other, but to be a good scientist you have to do both."

A "now famous" anecdote about Graham Pyke and RMBL has to do with his departure from Gothic in 1985. Not knowing when he would be returning, he asked David Inouye to hold on to the cooler filled with personal items that Pyke had been storing there each winter. "Over a twenty-two year period, David dutifully looked after my cooler and moved it from one place to another. When I came back in 2007, it was up in the attic in the lab. When we opened it, there was an unmistakable smell of Vegemite! It turned out that I had left a container of Vegemite, sent by my mother. Over the years, the freezing and thawing had cracked the plastic container she had put it in – it was nothing but a lump of concrete! I actually thought about trying to resurrect it, but in the end I just tossed it out." Such a shame… "David always insists Vegemite is bear repellant or axel grease, but it goes really well with eggs!"[4]

Over the decades since 1928, there have been other international scientists visiting and working at RMBL, but not many have maintained a forty-plus year record of association. Even when not in residence, the personal and professional affiliations forged there have resulted in collaborations, scientific articles and new areas of pursuit, continuing into the twenty-first century.

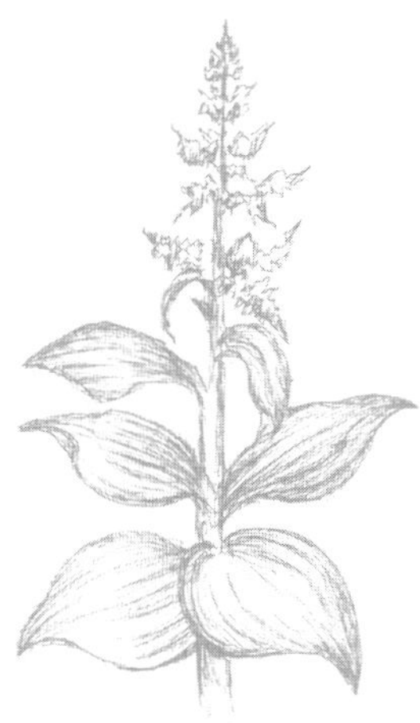

How to be a Good and Highly Cited Scientist According to Dr. Graham Pyke

Dr. Graham Pyke has published more than one hundred scientific articles, covering a wide range of topics and including a broad range of organisms. Many of these articles have been cited by other authors and in 2004 he was recognized as a "Highly Cited Author" in the category of Ecology/Environment.[1] This put him in the top 0.5% of researchers in terms of citations in this area and he mentors a program to help other people become better cited. Having one's research and publications frequently cited is important so that discoveries and discussions in the field of science, or any field of study for that matter, can be shared throughout the world. He has thought long and hard about what makes one a good scientist and a highly cited author, and these "secrets" are what he offers as the key.

Fundamentally, one must have a basic curiosity, a desire to understand the way the world works scientifically, as well as a keen desire to tell the world about what you find. "There are lots of people who are inclined in one direction or the other, but to be a good scientist you have to do both." Then follow the "secrets" of successful authorship.

"The first secret is you have to choose a scientific question that's of high, preferably immense significance, and it has to be achievable. You can't just pick something that is pie in the sky. It has to be really significant and achievable.

"The second thing is you have to have a passion, an absolute passion, for wanting to influence people. You want to change how people think, what they say and what they do.

"The third thing is, in the Australian vernacular, you have to tell a bloody good story! You have to tell a story that is both compelling and memorable. It has to be compelling because you are trying to influence people. You want them to become your disciples, so to speak, so you have to be good and you also have to make them remember where they got the idea from, so that when they do their work they will cite your paper in their article.

"These three qualities are, I think, measures of success in lots of different areas, not just science. You can just change the terminology slightly."

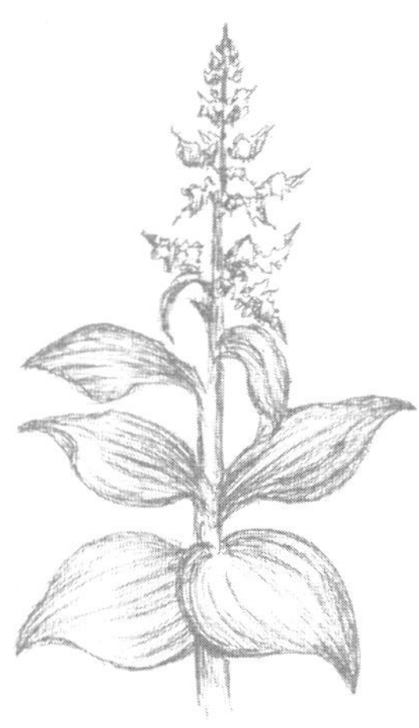

Challenging Dogma

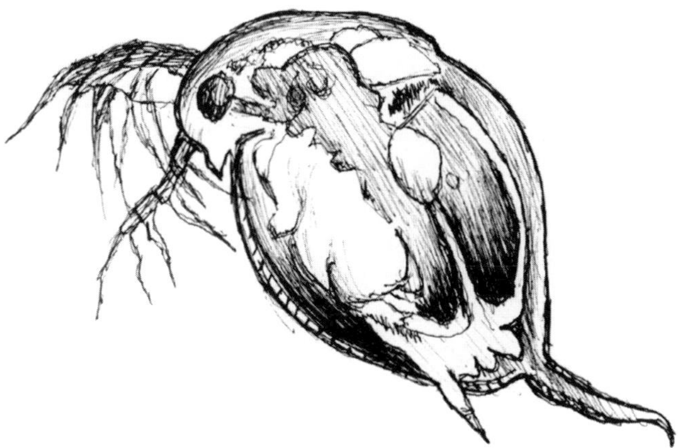

Dr. Stanley I. Dodson (1944-2009)

Not everyone who has made a mark at the Rocky Mountain Biological Laboratory was necessarily in residence for many summers. Some have used the forests, fields, ponds and streams offered by Nature for a number of years; others have passed through and influenced - or been influenced by – the students and investigators in a significant way. One individual who has been identified by more than one scientist as important in the story of RMBL is Dr. Stanley I. Dodson, Professor of Zoology for many years at the University of Wisconsin, Madison.

Born in Lincoln, Illinois, Dodson began his love affair with the deserts, mountains and ponds of the western United States at the age of ten when his family moved to Boulder, Colorado. He did his undergraduate work at Yale University, where he came under the spell of Charles Remington and Remington's graduate student, Ward Watt. There he also met his future wife and partner, Virginia (Ginny).

Stan Dodson died tragically in a bicycling accident in Colorado at the age of 65. His former student and life-long friend, Bobbi Peckarsky, has offered the details of his life and work at RMBL for this essay.

Young Stanley Dodson first experienced a summer at RMBL in 1963 when he was a freshman at Yale. Charles Remington and Ward Watt encouraged him in his efforts to gather information about "parthenogenetic lizards" in the lowlands of Colorado. At one point in the summer he was spotted collecting butterflies by RMBL researchers, Bob and Scottie Willey. Butterflies being subjects of study by both Remington and Watt, this observation does not seem too unusual. But Dodson's interests lay more in the area of water than terrestrial, and he followed his nose in that direction. He made his mark early, while still an undergraduate, co-authoring a classic paper on zooplankton with his professor, John Brooks. Interestingly, he managed to disprove one of the hypotheses formed in that paper when he settled into his doctoral dissertation research at the ponds of the Mexican Cut Preserve near Gothic.

Based on research on zooplankton done in lakes of Connecticut, the paper written by Brooks and Dodson established two hypotheses to explain "variation in size composition over space and time". One of these hypotheses was that "size-selective predation" (predators preferring prey of a certain size) by fish, specifically alewives, which feed on zooplankton, leads to the elimination of larger-bodied species of plankton. Secondly, "the size-efficiency hypothesis" was that larger-bodied species of zooplankton have a competitive advantage over smaller zooplankton where there are no alewives present. This second hypothesis is the one disproved by Dodson using the ponds at the Mexican Cut Preserve for his further study.

In the words of Bobbi Peckarsky, "From the beginning, [Stan] embraced the challenge of interpreting unexpected results and thereby generated new ideas".

Stan Dodson had an especially important impact at RMBL as a mentor. Dr. Theo Colborn met him at RMBL and was thrilled to be able to work with him on her PhD research. Peckarsky has credited him and his wife, Ginny, not only with her decision to study aquatic ecology, but also on her approach to science, education and life. Stan and Ginny taught courses in aquatic ecology in Gothic for four summers – 1974, 1978, 1979 and 1981. After that, their trips to RMBL were intermixed with family vacations to Grand Junction.

A distinguished and premier zooplankton ecologist, Dodson is remembered by many as much for his contributions to his field of study as for his impact on the growth and development of his students. Peckarsky shares a tale of Dodson preparing his General Ecology students for a field trip to the International Crane Foundation. He asked a Tai Chi instructor to lead his students and staff in learning and performing crane movements prior to visiting the cranes. "This example epitomizes a mainstay of Stanley's teaching philosophy – one should involve the whole person, not just the intellect, in science; and, science should intersect with our daily lives."

Dodson taught the importance of trusting one's observations. If they fly in the face of existing dogma, make new dogma. He also encouraged the practice of keeping an open mind at all times and paying attention to negative results, as well as positive ones.

Reminding his students that "very famous scientists operate under the same rules as beginning graduate students", Dodson can be said to summarize much of what RMBL is about: Experienced investigators mentoring curious students with enthusiasm and respect guide those students to become the seasoned investigators of tomorrow.

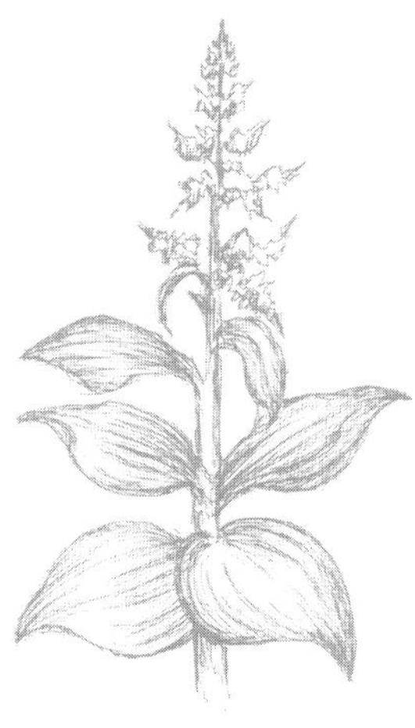

Stoneflies, Mayflies… Time Flies
The "Benthettes" and the Ecology of Streams

Dr. Bobbi Peckarsky

Benthic: "Of, relating to or occurring at the bottom of a body of water."[1] "Benthette": By extension, an individual who studies that which is of, relates to or occurs at the bottom of a body of water; especially if she (or he) works at the Rocky Mountain Biological Laboratory with Dr. Barbara (Bobbi) Peckarsky.

On a chilly afternoon in August, in an appropriately wet environment, surrounded by thunderstorms of increasing magnitude dumping rain at a more and more rapid pace to form impromptu streams down the hillside, Bobbi Peckarsky amiably chats of her passion for place and work. The place – Gothic, Colorado; the work – studying those tiny little creatures that live in the rivers and streams surrounding the place.

Peckarsky was a high school teacher in Wisconsin in 1973, spending summers enhancing her training and seeking to satisfy her scientific curiosity by taking courses at the University of Wisconsin. She spent some weeks each summer working as a waterfront director at a day camp in Madison, on one of the lakes. It so happened that many of the children of faculty at the university attended the camp. Her years working for bachelor's and master's degrees in zoology at the university, and her contacts with the parents of her summer charges, placed her in a position to "rag" on some of them about the paucity of summer courses. Having taken all that were offered, she asked that new ones be made available. One of the faculty suggested that she leave the area in summers and go to a field station. "I was just naïve enough to say, 'what's a field station?'" He enlightened her about two – Flathead Lake in Montana and the Rocky Mountain Biological Laboratory in Colorado.

Learning that Flathead Lake was a large facility, essentially the summer home of the University of Montana science departments, she investigated RMBL. Its smaller, more intimate environment was just what she was looking for, so she applied for the 1974 summer session. She was accepted, along with another teacher from her school and two students from the university. ("They pretty much took anyone who applied.") Three of the four drove out to Colorado together, camping and hiking along the way.

Peckarsky had always been an outdoors person, having been instilled with love of nature by her father. But summers in a cottage on a lake in Wisconsin were not the same as living in an ancient, run-down cabin in the mountains of Colorado. Oh-Be-Joyful Cabin had not, at that time, been renovated. At one time it had been a stables and the first floor was "just awful". One of her roommates, a mammologist, snap-trapped mice under their beds at night – "oh, yuck!" But she loved the fact there was a little window where she could look out from her cot upstairs; she could see the top of Gothic Mountain as the sun rose in the mornings. "I would know what time it was in the morning because the sun would hit the top of Gothic and move down, and I would hear the breakfast bell."

Although she was not yet enrolled in a Ph.D. program in 1974, Peckarsky was classified as a graduate student at RMBL. She enrolled in two classes that summer, 'Rocky Mountain Flora', taught by Harriet Barclay, and 'Aquatic Ecology', taught by Stan and Ginny Dodson. She loved the Rocky Mountain Flora class and felt particularly fortunate to be able to work with Harriet Barclay, who was 72 years old at the time.[2] Exciting and satisfying as that was, she became "smitten with streams" in the 'Aquatic Ecology' course. "Serendipitously, the instructors were from Wisconsin!"[3]

As a graduate student with a master's degree, Peckarsky was expected to complete an independent project. Enamored of streams and the bugs that lived in the streams – stoneflies and mayflies – she chose to do her independent project for Dodson's class. At the end of the summer, he suggested that she should consider returning to university to complete her Ph.D. She chose to enroll again at the University of Wisconsin and began her doctoral studies in the summer of 1975.

The Dodsons were both water ecologists, but they studied still water, not streams. Peckarsky was more interested in flowing water. She began turning over rocks in the East River and Copper Creek and found it really fascinating. "You turn over rocks and put some bugs in a pan and you just can't believe how much life there is in these streams. It's really fascinating!" She cannot recall if her fascination was due to an "ah-ha moment" or was gradually acquired, but she had always loved water. The speed of the flowing streams after the annual snow melt was a powerful enticement for study. "Part of it was the marvel of how the bugs do it. The water is running really fast, it goes up and down; there is snow cover and ice cover part of the year. How do these guys do it? How do they live? How do they survive? How do they get through their life cycles?"

So Peckarsky began her research in bodies of water that few had seriously focused on previously. Dodson and his colleagues were studying predator-prey competitive interactions, as well as the community ecology and how food webs were constructed. But, they were doing it in ponds – not in flowing water. Those researchers who were working in streams at the time were "mostly fixated on the physical nature of those streams – they've got flow and crazy rock rolling and disturbance". She may have chosen to study the consequences of all that 'rocking and rolling' had she not been introduced to water ecology through Dodson, whose predator-prey focus was the one she chose to pursue – but in streams.

The predators that lived in streams were the "big, cool stoneflies". Their prey were the little mayflies that also lived in the streams. The independent project she wrote in the summer of 1974, and which grabbed the attention of Stan Dodson, was about stonefly-mayfly interactions. "Fast forward forty years – I pursued the whole angle of how important predation is in streams and how it affects other aspects of streams." Her work was an uphill battle, unconventional for the time and questioned by a lot of reviewers who thought that she would not be able to manipulate the environment. She was determined to persevere, however, and believes her biggest contribution was to stick it out. Crediting her mentor with her work ethic, she recites it as: "Don't just do what everyone else is doing. If your results fly in the face of accepted dogma, excellent! Make new dogma!"

After a number of years of "making new dogma", Peckarsky expanded her horizons to include fish that live in the streams – specifically, trout, and the food base of the communities. She was fortunate to have students from other places, other countries – New Zealand, Spain, Sweden, Australia – who contributed to her work by adding differing perspectives to the study. "Essentially, we've contributed what we think is one of the most detailed knowledge bases of a community of stream organisms anywhere."[4]

The species of trout native to the Gothic area is cutthroat trout, but the majority of trout living in the streams and rivers has become brook trout, which were introduced from the eastern United States. What Peckarsky and her students began to focus on is that while some streams had trout, other streams had no fish at all. "We followed our nose to see what the differences might be between streams that have fish and streams that don't have fish." The "long journey" of ten or fifteen years started with an observation "that was really bizarre". That observation was that the trout streams had higher densities of their favorite prey, the swimming mayfly, than non-trout streams. The reason this was considered bizarre was that one would think, quite logically, that the streams that had predators would have fewer prey, because they were eaten. Once again, they went in search of new dogma.

Replacing the question she had worked with before, "Can mayflies smell stoneflies?" with another question, "Can mayflies smell trout?" she and her student investigators discovered that the answer is "yes". They can. "We learned that trout actually scare more prey than they eat. They change the behavior and they change the life histories of these [mayflies]." Experiments in collaboration with Dr. Maruxa Alvarex of Spain using "essence of trout" revealed that the mayflies responded to all the tested species of trout – brook, rainbow, cutthroat, brown, and "even the suckers". Further investigations, particularly in cooperation with Dr. Angus McIntosh from New Zealand, revealed that mayflies avoid trout predation by eating at night, when the visual fish cannot see them.[5]

Experiments continue, revealing new and interesting information about the *Baetis* (swimming) mayflies, and Peckarsky's enthusiasm for the research and the results is palpable. Her biggest concern after forty years of work on her projects in Gothic is that there is no heir apparent. Many of her students, although dedicated while working and studying at RMBL, have chosen to work elsewhere or not to stay in academia. Others have pursued interests outside the field of stream ecology. "They all have something going on. No one is coming back."

In addition to her personal research, Peckarsky has worked for almost forty years on an ongoing non-consumptive water rights project at the Lab, begun by Scottie and Bob Willey. Twice each summer, she visits eight sites to sample the water quality in the streams. "You can use bugs to indicate whether the streams change or not." She funds this project mostly herself and is equally concerned about the long term continuation of this project as of her stream ecology projects.

In the words of Professor McIntosh, Peckarsky's research is "set to contribute to more responsible management of water resources. Stream organisms are adapted to natural flow regimes, which are undergoing considerable modification due to increasing human demand for water."

Recent research has focused on "understanding how natural variation in flow influences the function of river ecosystems. It will contribute to prediction of the effects of altered flow regimes on river food webs… By working on the same study system, the East River drainage, for over 40 years (1974-present), [Peckarsky] has produced insights only possible through a dedicated 'place-driven approach to research."[6]

Having retired from Cornell University in 2005, she continues her research annually at RMBL as professor emeritus and adjunct professor at the University of Wisconsin. In 2008, she received the award of excellence in benthic science from the North American Benthological Society. She hopes to continue working for many years, citing Harriet Barclay as her role model. Her hope is that, "somebody will carry the torch after I can't do it anymore".

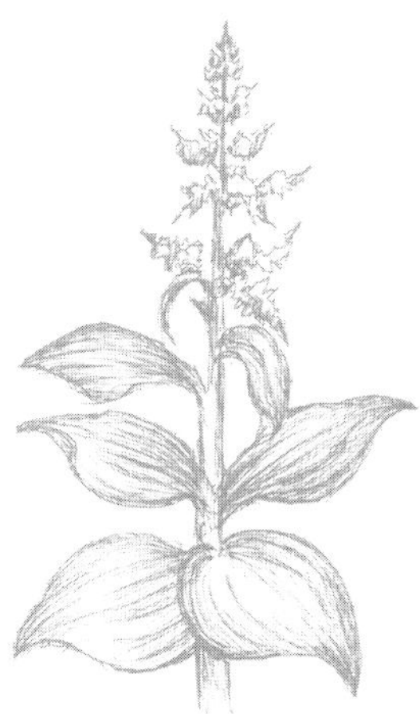

Meet the Burying Beetles

Dr. Rosemary Smith

If you have hiked in the mountains of Colorado, or at least have taken more than a few strolls through mountain paths in the area of Crested Butte and Gothic, you may have asked yourself, 'Why are there no dead mice around here?' Or maybe not – it is, in general, not in the mindset of hikers to look for dead animals, let alone dead rodents. It is much more exciting to catch a glimpse of something living. This question, however, has caught the attention of Dr. Rosemary Smith, who has studied the little insects that are a large part of the explanation for the rarity of dead rodent sightings.

Rosemary Smith arrived in Gothic in 1983, responding to her professor Paulette Bierzychudek's advertisement for a field assistant. While in Gothic, she worked for her professor and used the money she made to take a class with Steve Nowicki, a prominent ornithologist from Duke University. She so enjoyed the summer that she returned the following year, after her graduation from Pomona College in Claremont, California. Continuing on to graduate school at the University of Arizona, she "finagled" ways to return to Gothic each summer, working for RMBL or for her advisor, Michael Rosenzweig, who ran a project studying bee foraging behaviors. At last, as a Ph.D. candidate, she "invented a research project that would have to happen at RMBL". That particular project was focused on chipmunks and ground squirrels. She completed her graduate studies in 1991, returned to Gothic to teach a field course in Behavioral Ecology - and met her future husband Rick Williams, a botanist, on a climb and hike on Baldy Mountain that summer. They soon became annual regulars at RMBL, each pursuing their own particular research project. They married in a field outside Gothic in June of 1997.

Rosemary introduced her "burying beetles" (genus *Nicrophorus*) in an article she wrote for the 2004 Rocky Mountain Biological Laboratory *Autumn Newsletter* as follows:

"Someday you ought to put a small dead rodent outside, wait and watch. Within minutes flies will hover about and ants will wander by but, if you're patient, a surprisingly handsome, large and clearly purposeful beetle will fly in, land and begin to investigate the corpse. He (or equally likely, she) will walk over and under the dead mouse, heave it up into the air, poke and prod and actually come to a decision."[1]

The decision the adult beetle has to make is whether or not the rodent – usually a mouse or a vole – is suitable as a source of nutrition to nurture the growth of its offspring. The beetle may decide it does not like this particular rodent, in which case it simply goes on its way; or, it may decide that it is perfect for its purposes - in which case it must move quickly to establish its claim.

The particular beetles that Smith works with play an ecological role as decomposers. They take the carcasses of dead animals and use them to nurture their offspring. They do this by burying the carcasses and, after suitable preparation, laying their eggs nearby. The carcass then becomes the food resource for the developing young. Smith is particularly interested in two angles of this behavior. The first is, what ecological role do these beetles play? Secondly, she is interested in the behavior of the parental beetles.

In an area the size of a football field, Smith says that there might very possibly be between fifty and two hundred living mice. As their life span is approximately two years, in any given year approximately half of the population will die. That is a significant number of dead mice! Naturally, some are found by scavengers such as coyotes or weasels Smith has found that flies and ants usually arrive first and work diligently to deposit their eggs on the carcass. So, time is of the essence when a burying beetle is introduced into the story. A mouse or other small rodent that dies at nine o'clock in the morning will, by one o'clock in the afternoon have been assessed and buried under ground, out of sight and away from other scavengers, by the beetles – assuming it is an acceptable specimen. The question that arises, and by which Dr. Smith is intrigued, is – just how important are these decomposers in ridding the surface of the earth of dead animals?

It appears that they are quite efficient at this. The mated pair of beetles is perfectly capable of excavating the soil near the carcass and digging a hole in which to deposit it. Having dug the hole, they use their bulldozer-like legs to push the soil until it causes the mouse to fall into the hole. However, it is much more desirable – faster and safer – to use a pre-existing hole. The beetles will search an area perhaps three to five feet away from the dead mouse. If they find an acceptable hole, such as a gopher hole, they will drag the carcass to the edge of the hole and push it down. Now for the "yuck" factor…

Having successfully chosen their mouse, the male and female prepare the dead mouse for the final role in its existence. First, they remove the fur. Then, they roll the mouse into a ball and pack the soil so as to form a small chamber. They then mate and the female deposits the eggs into the soil surrounding the mouse. Within a day, the eggs hatch. The parents make sounds with their wings that attract the hatched larvae to the mouse. They then feed those larvae with a liquid, derived from the carcass, until the larvae are old enough to feed on their own. They will continue to do so until the entire carcass is consumed, which takes about two weeks. Interestingly, the parents remain with the offspring until the process is nearly completed, after which they fly off separately to find a new mouse and, if they are lucky, begin the process all over again. The larvae remain underground until the following summer, when they emerge as adults.[2]

This brings the discussion to Smith's second area of interest, which is the behavior of the adult, parental beetles. Why the parents remain to oversee the consumption of the mouse is not known. It has been observed that one or both of the parent beetles can abandon the site and it has no impact on the growth or survival of the offspring. So, why do they stay? This question is under ongoing study.

Another interesting fact is that the parental beetles actually cull their offspring. In general, each mating results in around twenty to twenty-five eggs being laid. However, most likely only about ten or twelve of those offspring will be reared to maturity. Why does this happen? It has been shown that the parents are able to determine the size of the mouse and cull their brood to the number that would most benefit from it. According to Smith, "they can weigh [the mouse] while they are rolling it up and determine if it is, say, a twenty-kid ball. Or, it might only be a ten-kid ball, or an eight-kid ball. As they always lay an overabundance of eggs, they bite off the heads of the ones they don't want as the larvae grow. This means they are a bit more calculating than we thought. They are able to weigh something perceptively and count their larvae, or at least have some sense of how many larvae will survive a mouse."

Smith's research has shown that there are fitness consequences to the number and size of larvae. If there are many offspring, they are likely to emerge as small adults - if they even survive the winter. A small adult will not be able to compete with larger adults for mice. However, if fewer, larger larvae are allowed to survive, they might be large but their small numbers mean that the parents' contribution to the next generation is not very great. Therefore, "parental decisions are a finely tuned balance between number and size, adjusted to the resources available".[3] The tuning, of course, is done by Nature itself, selecting for successful parental strategies that leave them with the optimal number and size of descendants.

It was once believed that the adult beetles lived only one season. However, trapped animals that were marked one year by the researchers have been spotted the next year, indicating that at least some of them survive. They, as well as the larvae, burrow in the soil from August until the following June. They all live off their fat – the larger the beetle, the more likely it will survive in the relatively shallow compartment it has created in the soil.

Rosemary Smith is a great believer in what she calls "curiosity-driven" research. Such research helps us to understand how life works and can lead to new discoveries by some future scientist. It need not have any acknowledged practical application purpose. Curiosity-driven research "that has absolutely NO practical application whatsoever is part of what makes us human. It is like music, like the arts, like sports; they do not make us healthier, they don't do anything. But they are what make us human, make us civilized… There is no actual benefit of studying beetles per se, but it is super cool. I feel fortunate that I get to be a scientist and discover stuff. Even if it is just cool."

Smith's positive outlook concerning the importance of "fun" in curiosity-driven science is complemented by a series of discoveries that may have significant practical application value. One has to do with discovery of the anti-bacterial and anti-fungal properties in the secretions with which the beetles coat the rodent carcass when preparing it for the offspring. She and a master's student of hers worked to identify the substance in the saliva which prevents the growth of bacteria and fungi. Such a substance may be of particular interest to people in the food industry, especially the meat industry, since it may prevent the growth of bacteria and fungi on meat stored at room temperature. "Of course," she says, "you would not put beetle spit on food for people!" But just as with antibiotics, a product discovered from an organism might eventually be synthesized in a laboratory. As it turns out, Dr. Smith has looked into applying for a patent on her discovery of the anti-microbial, anti-bacterial and anti-fungal properties of beetle spit.

Another intriguing possible application for beetle spit is as a depilatory. The beetles' actions to remove the fur from the rodents they are preparing as a food source is not un-similar to a natural chemical present in animals that molt. It causes their hair to fall out. So, the question is, can the property in beetle spit (or perhaps that of other animals) related to hair removal be synthesized into a depilatory safe for use by humans?

The research on burying beetles continues, with new generations of master's students working with Rosemary Smith to pursue particular questions of study. In the meantime, the beetles continue to search for dead rodents, evaluate their size and potential value, bury them, prep them, lay their eggs in them, cull their offspring and watch over the ones they have chosen to survive.

When next you are hiking and see the carcass of a mouse or other rodent, stop for a while, have a drink of water, enjoy the surroundings – and keep an eye out for a good-looking beetle on the prowl for a dead mouse.

Supersized Chipmunks? No, Golden-Mantled Ground Squirrels

Dr. Rosemary Smith

Dr. Dirk Van Vuren

Chipmunks are 'special' in that most people would agree they are "cute" rodents. They proliferate in urban areas, where they raid bird feeders for seed, as well as in rural regions where flowers and grasses provide the sustenance they need to survive. Pesky but adorable, almost anyone would recognize them at a glance. In and around Gothic, Colorado, however, there is a somewhat larger version of what many would consider to be an overgrown chipmunk – a supersized chipmunk. This animal is not a chipmunk at all, but a golden-mantled ground squirrel.

Dr. Rosemary Smith of Idaho State University focused her first research project in Gothic on chipmunks and golden-mantled ground squirrels in the 1980s. An animal behaviorist, she chose to work in Gothic because she was interested in how predators influence the behavior of animals – and because she loved it there! She moved on from studying squirrels to beetles, but maintains an ongoing interest in the continuing research being done on ground squirrels.[1]

Dr. Dirk Van Vuren spent his first summer in Gothic in 1982, at the invitation of Dr. Ken Armitage. Van Vuren was to begin his graduate studies at the University of Kansas and Armitage asked him to trap yellow-bellied marmots for his research study. That experience motived him to make marmot dispersal the focus of his doctoral research – he became a *Marmoteer*! He received his Ph.D. in 1990 and moved a month later to teach at the University of California, Davis. Continuing his work on marmot dispersal for another five years, he developed an interest along the way in why some animals are social and live in groups, like the marmots, and why some are asocial, like the ground squirrels. Accordingly, in 1990 he began to study ground squirrels, and continues to do so. He spent a total of 14 years doing marmot work and 25 years focusing on squirrel work, six of those years overlapping. His graduate student, Caitlin Wells, continues his research on ground squirrels.

Golden-mantled ground squirrels (*Callospermophilus lateralis*) are smallish squirrels, about eight to twelve inches long, that live in a variety of habitats in mountainous areas. They have stripes the length of their bodies, but not on their heads, and have a white ring around the eyes. Their tails are not as long or fluffy as those of a tree squirrel, or a chipmunk. They do not hold their tails up or flip them, as might a chipmunk or tree squirrel. Barring these differences, it is easy for a casual observer to mistake the ground squirrel for a chipmunk, especially at a distance.

According to Smith, the risk of predation is a primary factor in determining where the golden-mantled ground squirrels live and how they co-exist with the smaller chipmunk. Both burrow in the ground and hibernate, or rest over the winter, as do marmots, and both prefer rocky areas that make it difficult for predators, including foxes, coyotes and weasels, to access their burrows. They also have very similar diets. Due to their small size, foraging in open spaces is an invitation to danger, so living areas must be carefully chosen to reflect the need for protection and the availability of good food resources. Smith found that it was this balance that explains the coexistence of the two species: "The larger golden-mantled ground squirrel runs faster and is more vigilant than the smaller chipmunk, so it can exploit the riskier open meadow habitats. But, the smaller chipmunk is a more efficient forager, so it can gather up all the good food nearer the rock piles and burrows. By dividing up the habitat based on their different perceptions of risk, they can both make a living in the same environment."

As with marmots, male dispersal is part of golden-mantled ground squirrel behavior, meaning the males leave home early and the females stay around the burrow. Females do occasionally disperse, but mostly remain close to their burrow. The adult ground squirrels emerge from hibernation in April and/or May, depending on the year. Mates are chosen or accepted and mating occurs almost immediately after emergence. Females have only one litter a season and gestation takes about thirty days. The average litter size is four, but some older females have been known to produce six or seven pups. Most pups are born in June. For the first month or so after birth, the pups remain in the nest being nursed by their mother. Once they are weaned, they quickly become hungry and start emerging from the nest to forage. This "date of emergence" is when the researchers see the pups for the first time. Since they are underground for the first month of their lives, the investigators cannot determine the size of pups at birth. At the date of emergence, however, their weight averages between 50 and 70 grams (1.75-2.45 ounces); the mothers weigh between 160 and 250 grams (5.65-8.82 ounces) at that time. By the end of summer, the juveniles quadruple their weight.

The date of emergence from hibernation varies from year to year. It is related to the amount of snowfall and how long the snowpack lasts. This information is readily available at the Rocky Mountain Biological Laboratory. Since the 1970s, billy barr has spent his winters in Gothic measuring snowfall, recording temperatures and other weather data. Thanks to billy, it is not necessary for investigators to be present to measure the snowpack from December to April; they need only to look at the information on the RMBL website or arrive in the spring and check the records.

Small metal traps with a one-way door, called Tomahawk traps, are used to capture the little animals. Peanut butter and sunflower seeds are always used together as bait. Sheeting is utilized to shade the tops of the traps so as not to cause the creatures heat stress. Once captured, they are weighed and sexed, their ears are tagged and they are marked with a black dye, such as is used to mark the marmots. One can always tell a "marmoteer" or ground squirrel researcher by the black stains on his or her hands.

Ground squirrels are diurnal animals, which means they are awake and about during the day. This benefits them because many of their predators stalk at night; but, on the other hand, they do not have good self-cooling mechanisms. In response to this, they frequently re-enter their burrows during the heat of the day, where it is cooler and moister. They tend to be most active from about 7:30 until 10:30 in the mornings, and again from about 3:00 until 6:00 in the afternoons. They eat seeds – thus their scientific name *Callospermophilus*, which means "Beautiful seed lover". Seeds not only provide a lot of fat, essential for their survival during hibernation, but also most of their water. Although they are not often seen drinking water, it is not uncommon for them to sit next to their burrows and eat snow when they emerge from hibernation. It is a quick way for them to obtain necessary water before there is an abundance of food.

Just how long ground squirrels live is a tricky question. As with many mammals, the first year is the key to survival. Many of them do not emerge from hibernation or, if they do, do not make it through their first summer, largely due to predation. That being said, however, the researchers have documented females that have survived five, seven and even nine years. Sounding kind of proud, Caitlin Wells described the oldest female identified: "Our nine year old female was a champion; she had a big litter each year. But, she got into a lot of fights so her ear tags would pull out all the time. We kept replacing them until her ears were only stubs and we couldn't get any more tags in. So, she just became known as 'Earless'!"

One of the worst hazards to the survival of these little animals is human lack of awareness and carelessness. Dogs were banned in RMBL decades ago. Cars, in particular, driving through the center of Gothic at speeds never meant for the roads, are the cause of many lamented animal deaths. The road through Gothic is a vestige from mining days and cuts the town neatly in two. Anyone privileged to spend summers in the town will, given the opportunity, launch into a rant about the evils of the automobile in a biological field station. The town area was designated a School Zone and posted with the associated speed limit of 15 mph, something that is all too often ignored. One of the researchers thought to install a "marmot crossing" sign, but the county objected because it wasn't an "official" sign. Someone came upon the idea of acquiring and parking a cast-off, non-functioning old sheriff's car, with official lights on top, along the roadside. For a while, it was parked partially in the shrubs and bushes and a "dummy" was placed behind the wheel. Unsuspecting motorists obligingly hit the brakes in fear of the long arm of the law. The annual arrival of an increasing number of visitors to both Gothic and the surrounding mountain lakes and trails is a mixed blessing. People are welcome; slow speeds and vigilance are the keys.

Caitlin Wells has focused part of her research on the impact of an invasive ground squirrel from northwest and central Colorado on the indigenous golden-mantled ground squirrels. These 'Wyoming ground squirrels' were first seen in the 1990s in the Almont Triangle area, between Crested Butte and Almont. In 2010, Wells trapped and tagged one of these invaders in Gothic. Speculation was that perhaps it had hitched a ride on a vehicle coming into the area from its original habitat. This is known as "human mediated dispersal" – traveling to a new destination via means provided by humans, rather than migrating on foot. Physically, the "Wyomings" are not as adorable as golden-mantled ground squirrels, more closely resembling the much maligned prairie dogs, only smaller. According to Wells, the Wyomings existed in the area at one time, but disappeared during a long drought. It appears they have returned. In 2011, Wells found that a Wyoming had dug a burrow and hibernated in Gothic.

The first Wyomings to show up have been only males but, according to Dirk Van Vuren, that is only natural. The young males disperse and spread out in search of new, friendly territory. They settle in and, within a few years, the females start to arrive. At that point, he says, "All hell breaks loose".[2] Since Wyoming ground squirrels have larger litters than the golden-mantled variety, they can take over a field fairly rapidly. This makes them unwelcome to the indigenous population, as well as to ranchers and farmers.[3]

"Interestingly", comments Van Vuren, "it is not known why Wyoming ground squirrels displace golden-mantled ground squirrels." Wyomings are a little larger than the golden-mantles ones, and it is guessed that they are more aggressive. "But no one knows. We hope to find out, by being on hand to witness the colonization of Wyoming ground squirrels and the fate of golden-mantled ground squirrels – no one has done this before."

According to Rosemary Smith, a major concern about the invasion of the Wyomings, especially in the area around Almont, is that they may threaten the Gunnison sage grouse. These ground squirrels have been known to prey on eggs of ground-nesting birds, of which the sage grouse is one. Some studies have shown that the mouths of the Wyomings are too small to eat a sage grouse egg, but they will eat a broken egg. Careful observation over time will be necessary to learn more about this potential threat.

Meanwhile, Wyomings have been seen to take over dumpsters in Gothic, which are a rich source of food for ground squirrels, as well as a safe haven from predators. The Wyomings are more aggressive than the golden-mantled squirrels and pose serious competition to them. Only time will tell if the golden-mantled grown squirrels remain in Gothic, or if they are chased off and superseded by the Wyomings. However it plays out, they will be watched by researchers!

Botany for the Ages

Dr. Charles F. "Rick" Williams

Intergenerational studies in any field of academic inquiry are as valuable as they are unusual. Frequently, a student pursuing a Ph.D. will complete his or her research, possibly based on something that was available and could be investigated during the prescribed five years that are allowed to complete a dissertation, only to put it on the shelf for that very reason. An investigation into what may have appeared to be an intriguing question might lead to answers, but also to loss of interest in the subject matter. And, even if the researcher chooses to continue in the same vein, there is no guarantee that the material will be of interest to someone in the next generation. End of subject.

Enter people like Charles F. Williams, known to friends and colleagues as "Rick". His interest in plants and flowers began early in his life, while growing up in Tulsa, Oklahoma. It was, one might say, in his blood. His father operated a commercial greenhouse and his mother was an artist. A graduate of the University of Tulsa, his mother had taken a class in botany during the 1940s, taught by Harriet Barclay. One of her classmates was Jean Langenheim – quite an introduction to the powerful distaff side of the Rocky Mountain Biological Laboratory.[1] Rick got to know Barclay when she spoke to his classes in the Tulsa public school system. He worked several summers at the Red Bud Valley Nature Preserve in Tulsa, which Barclay had started with help from the Nature Conservancy. Having completed his first year of college at the University of Oklahoma in 1977, he finally followed Barclay to Gothic, Colorado and the Rocky Mountain Biological Laboratory, a place he had heard much about. There, he took her flora class and a mammal ecology class, taught by Dr. David Armstrong from the University of Colorado, Boulder.

Williams' graduate work led him in different directions and he did not return to Gothic again until 1991, when he had completed his doctoral dissertation. There, he met (and later married) Rosemary Smith. The two have returned to Gothic every summer since then.

The focus of Williams' work is the genetic aspects of plant reproduction, "mostly plant mating systems, how pollinators interact with plants and flowers, and how that influences whether flowers and plants self-fertilize and out-cross fertilize", as well as how that influences the fitness of the plants. Many of his experiments are "manipulative", where he prevents certain kinds of mating from occurring. He looks for genetic markers in the plants and seeds to determine paternity. Using this information, he works to determine what plant characteristics influence success in mating, who the parents actually are and the fitness of the "out-cross" versus inbred offspring. From there, he tries to determine how all this "influences the evolution of the plant traits that determine all of the mating patterns in plants." He works with native plants, particularly larkspur and wild geraniums.

Together with Dr. Nick Waser from the University of California, Riverside, Rick conducted a "fairly long-term study" to investigate some predictions made based on Waser's pollination work over the previous twenty years and attempted to support those predictions. Becoming focused on pollination as opposed to his earlier interest in seed dispersal, Williams started comparing pollination systems of related species. For example, there are two species of larkspur – one that blooms right after the snow melts and is small compared to the other, a big, tall variety that blooms later in the summer. He was concerned with such questions as how the variation in plant size influences pollinator behavior, and how it might influence mating patterns. This became the focus of his continuing work.

Since moving to Idaho State University in 1999, Williams has worn two hats. One is as Associate Professor of Biology; the other is Curator of Botany at the Ray J. Davis Herbarium, Idaho Museum of Natural History. It is in this second position that he began compiling, organizing and documenting information on plant collections and creating an online database. The purpose of such a database is not only to record information about the plants, but to enable people to work with the information without having to travel to a particular university or museum. Williams and his colleagues not only record the information, but also photograph all the specimens so that a researcher can actually see it and measure it and describe it and do whatever he or she needs to do with it.

Dr. Williams is not alone in his efforts at the Idaho Museum of Natural History. He works with groups that are documenting all the plant collections in the American Great Basin and the Pacific Northwest. The collection he personally is in charge of has about 80,000 specimens, but is relatively small compared to those of the bigger museums at universities like Brigham Young, the University of Utah, the University of Washington and the University of Oregon. Together, these universities have millions of specimens. All of these places pool their data in an attempt to provide a one stop data base containing known information about the plants of the regions.[2]

Harking back to Harriet Barclay's years in Gothic, the field station continues to maintain the herbarium that she began. A herbarium is a dried plant reference collection. Barclay's collection dates back to 1927 and has been stored in files in Gothic. The cold and dryness of the climate are ideal conditions for preserving plant specimens. The Gothic collection contains only about five thousand specimens and, happily for Williams, most of the data related to these specimens were entered into the computer data base one winter by George Aldridge, a winter caretaker in Gothic. Williams has worked "to clean it up" and add it to the big database he is creating. With help from students and volunteers, he has photographed all the specimens and uploaded that information. The next step has been to update information about each specimen, such as its exact locality and put it into a map-able format.

Truly fascinating is the process used to preserve plant specimens in the twenty-first century. To start, nothing appears to have changed from what the average child would do to "save" a lucky four leaf clover found in the yard. A specimen is selected, information about it is recorded and it is pressed and dried between sheets of newspaper. The specimen is then physically entered into the herbarium collection. Where the process diverges from tradition is that the information is then entered into the digital online database, where it is available for anyone to use.[3]

Not only did Harriet Barclay contribute to the RMBL herbarium, but her student Jean Langenheim completed a landmark vegetation study of the Gothic area, including the East River valley, Brush Creek and

Cement Creek. Langenheim's PhD dissertation entitled, "Plant-Ecological Reconnaissance of the Crested Butte Area, Gunnison County, Colorado", was donated by her to the RMBL archives.[4] She collected specimens and described the disposition and composition of the plant communities in over 200 sites, across one hundred square miles. As far as is known, no one had returned to resample those areas until Williams and his graduate student, Stephanie Zorio, began a project to do so in 2012. Zorio has been using RMBL records to locate sites "usually far off trail, in the most beautiful places you can imagine".[5] Once a site is located, she sets up a transect line three hundred meters long. Every meter, she records what she sees and makes careful note of any plants that appear to differ from what Langenheim recorded. According to Williams, "We are starting to see some differences in the vegetation community structure in the sites and hope to be able to show that there have been directional shifts in the plant community composition over the 70 year period [since the original data], and then try to infer what the causal mechanisms for those changes might be".[6]

Jean Langenheim's Master's thesis investigated plant succession following a major landslide or earthflow below Gothic that occurred around 1923. Langenheim staked out the location of ten permanent plots on the earthflow. Beginning in 1947, she documented the plants present in these plots, as well as in many other sites in the earthflow. She then followed the changes in plant species for the next several summers. These plots have been resurveyed about every 20 years since Langenheim's original study by various RMBL students and researchers. One of William's undergraduate students, Lydia Paradiso completed another project to reestablish those plots and investigate how the diversity of plants has changed over time. Any plants that are not easily identifiable are taken to the herbarium at Gothic and compared with specimens deposited there, using modern technological means. Once again, RMBL investigators are in it for the long run.

Coming full circle from the early days of RMBL, the work of Harriet Barclay, family friend and academic mentor, continues to live on in the work of Rick Williams at the Rocky Mountain Biological Laboratory - and elsewhere. Specimens that were collected almost a hundred years ago have been physically preserved for posterity, and will now be preserved for history in the massive data base Dr. Williams is helping to create.

Fairy Trumpeting

Dr. Diane Campbell

There is no reason that one should be aware that fairy trumpets, or scarlet gilia, are one of the longest studied species of flower at RMBL. But they are. These delicate little reddish-orangey wildflowers that look just like their name - trumpets - have been the focus of research since 1984 by Dr. Diane Campbell of the University of California, Irvine.

Diane already had her Ph.D. and was doing post-doctoral work with Dr. Nick Waser when he encouraged her to visit the Rocky Mountain Biological Laboratory. Originally intending to study marine biology, she had become interested in plant populations and pollination some time during her first year in graduate school. While simultaneously doing a project on pollination for a class and attempting a project on aquatic biology as a dissertation idea, she came to realize that the pollination work was going well and the other was not. "That's when I saw the light!"

In terms of finding something to study, Campbell says that "you can approach field ecology in two ways. You can either start with a question and look for an organism, or you can start with an organism in a place and look for a question." In her case, she began with a question and was looking for the right plant with which to pursue it. Her interest was in the evolution of flower morphology, which is the study of the form and structure of an organism. Although the subject had been studied since before Darwin, the part of it that was of interest to Diane was that most flowering plants are 'hermaphrodites'. They reproduce as both male and female – the male side makes pollen, the female side makes seeds. Research done prior to 1984 on plant reproduction had focused mainly on such subjects as how seed production was related to flower size or flower color. "But, since in principle they are doing half their reproducing as males, that leaves open the question whether the traits that generate high seed production are the same traits that would make the plant successful at dispersing pollen to other individuals." She wanted to look at that side of the question.

Needing a plant that had interesting variations in flower traits and on which she could track the pollen, she chose to follow the advice of Dr. Nick Waser and Dr. Mary Price and look at the scarlet gilia. They had done a study a few years earlier in which they used fluorescent powder dyes to color the pollen grains of the scarlet gilia. This made it possible to look at stigmas on the female reproductive surfaces of other plants to see how much of the dye had been dispersed. This was a method Diane could utilize to track how successful a plant was at dispersing pollen. So, she began her work on scarlet gilia in 1984 and has been working on plants in the same family ever since, as well as extending her questions to other areas of interest.

For the most part, scarlet gilia are pollinated by hummingbirds, with about five percent of the work being done by insects such as butterflies, hawk moths and long-tongued bees. What she learned from her initial study is that seed production is only part of the story. Some flower traits are selected only because of their effect on male function. One of those traits is the width of the corolla tube, formed by fusion of the petals and surrounding the anthers, which contain the pollen for dispersal. "If the flower has a broad tube, hummingbirds can insert their bill farther into the tube, they pick up more pollen onto their heads and are more successful at exporting that pollen onto other individuals." As for the female traits, the position of the stigma, "which is the female reproductive surface, whether it's down in the tube, at the level of the opening or sticks out, matters as to how much pollen is received by the plant and how many seeds it ultimately makes".

This particular project was what got Diane to Gothic and started on her journey with fairy trumpets, but her general interest is in how evolution works in natural populations. Therefore, she has been working in the field to study the processes of natural selection and movement of genes in populations in those fields. One project that has been her focus has to do with hybrids and what maintains them in nature. "These things don't seem to turn out as simple as everyone thinks." Everyone in plant evolutionary biology, that is.

A simple explanation for hybrids is they are not very fit, but are continually being made because of the movement of genes between species by pollination. Hybrids in nature generally don't survive very well; they do not reproduce. What she has learned, however, is that this is not always the case. In some instances, "the fitness of hybrids is strongly dependent on the environment. And, in this case, the hybrids between these two species [of Ipomopsis] can be even more fit than both of their parents, if they are growing in the right environment." They fare quite well in areas with dry soil and are quite efficient at using water, "especially if slender tube skyrocket is the seed parent, rather than scarlet gilia".

The paintbrushes in the Gothic area – those spikey pink, red and yellow wildflowers that carpet the high country – are a great case in point. They don't just hybridize with two species, but with four! One is pink, or magenta; another is yellow. Then, there are two different red species. "They can all make hybrids, which is why you find really cool arrays of color combinations." When it comes to columbine, the two most common in the area are the Colorado blue, seen many places, and the red, which are found in the woods. They are two separate species and therefore rarely, if ever, hybridize. There are, however, a variety of colors among the blue columbine, going from mostly white to mostly blue.

So why does it matter if a plant can and does naturally hybridize? Crop plants are hybridized by growers for any number of reasons, such as to make them more resilient to the cold or to drought or to insects. But in nature, this happens without interference by man. What Campbell finds interesting is that natural hybridization "reflects some breakdown of the mechanisms that keep two species apart. It tells us a lot about the evolution of new species. In regard to agriculture and conservation, many crop plants hybridize with natives – in fact, out of the thirteen most common crops grown worldwide, twelve hybridized with a native. It is important to understand the circumstances under which hybrids are successful and can survive and reproduce in nature."

Dr. Campbell has returned every year since 1984 to pursue her work with scarlet gilia. Four years she also traveled to alpine areas in New Zealand to do field research, which contrasted in interesting ways with her work in Gothic. "Up here, I can use what is now a well-studied plant to address certain fundamental questions where you really need a lot of background in order to push the science that much farther. The research in New Zealand was of an entirely different sort. From the first day I went out there, I was finding out new things that nobody had seen before, just because nothing on pollination was being done."

As for working at RMBL, Diane says that there are definitely changes about the place in general, but the kind of science has most certainly changed. "People do a lot more with equipment, electronics and so forth. The use of modern technology – that has certainly changed the nature of the science. When I compare when I first came out here in 1984, I had my paper notebook and my pencil and some string, some flags, measuring tape, some little labels to put on plants, those powder dyes that I applied, toothpicks to apply them with, little envelopes to put flowers in and I used a microscope. Now, an awful lot of the data I take is taken through some electronic gadget that is connected to my computer. So, if I measure the color of a flower, I am doing it with a reflecting spectrometer that is connected to my computer. If I measure photosynthetic rate, I'm doing it with an infrared gas analyzer that's automatically recording carbon dioxide and water coming out of the leaf, and it's connected to my computer. It's a little bit different way of doing science."

Three decades into it, the science may have changed but Nature's laboratory – the fields and streams and mountains – that surround Gothic continues to offer fodder for study. And Diane Campbell continues to return.

Climate Change and RuMBLers

January 2015 was, according to the National Oceanographic Atmospheric Administration (NOAA), the second warmest January on record; records date back to 1880. Only January 2007 ranked higher, and January 2014 did not lag far behind. "Average temperature across global land and ocean surfaces was 1.39°F (0.77°C) above the 20th century average. During January, the globally-averaged land surface temperature was 2.57°F (1.43°C) above the 20th century average… the globally-averaged sea surface temperature was 0.95°F (0.53°C) above the 20th century average.[1] Continuing the trend, March 2015 has the same distinction as January – it was the warmest March since record-keeping began in 1880; and, in the first quarter of 2015, "the average temperature across global land and ocean surfaces was 1.48°F (0.82°C) above the 20th century average. This was the highest for January–March in the 1880–2015 record".[2]

This is compelling data going toward proving the relentless progress of global warming or, more accurately, climate change.

There exist many naysayers when it comes to climate change, but it is rare to find one within the scientific community. And the community of scientists at the Rocky Mountain Biological Laboratory is no exception.

As pointed out in the essay about Dr. John Harte and his work, he began and, for twenty-five years in Gothic, Colorado, ran an epic study about climate change – proving that it is indeed occurring, and that it is having unanticipated and potentially unpredictable impacts on the flora of the West Elk Mountains. John and Mel Harte work diligently to educate the public about climate change through the popular media and elsewhere.[2]

Dr. Ken Armitage has written about climate change and marmots. Although conservation of marmots was traditionally impacted by human activities, such as hunting and threats to their habitats, regulation of those activities "reduced the threat of extinction". However, "Climate change, an indirect human impact, threatens marmot survival through global warming and extreme weather events."[3] Drought, changes in snow pack and temperature extremes all take their toll on marmot populations.

For four decades, Dr. David Inouye has kept track of wildflower blooming schedules in the mountains around Crested Butte. This seemingly small study has demonstrated that, during that forty-year time period, more than two-thirds of those flowers changed their blooming patterns in response to climate change. Many begin to bloom earlier, others reach their peak bloom earlier and still others bloom later into the year. Averaged over time, the flowering season is about a month longer than it was in 1975.[4]

These professional scientists are not alone in their discoveries, which abound throughout the work of researchers of all specialties at RMBL.

Climate change is controversial. It is discussed in the media, in political and economic arenas, schools, churches, around dinner tables. It is not within the realm of history to predict the future – but George Santayana's oft repeated maxim, "Those who do not remember the past are condemned to repeat it", is significant here. If the work of the scientists who dedicate themselves to proving what could well be a global catastrophe – the generalized warming of the planet – is ignored, what will become of this planet we call home?

Lost in Time

As with the mythical place called Brigadoon, so does the Rocky Mountain Biological Laboratory in little Gothic, Colorado disappear each fall and winter into the mists of time. It will emerge again, year after year, when the promise of spring and summer urges it and all that surrounds it and encompasses it back to life. But sleep it must; the plants, the trees, the wildlife and all living things that inhabit the biosphere of the West Elk Mountains must sleep, in order to rest, recuperate and regenerate. Researchers return to the lives they live in the "real" world. Information gets processed. Papers are written. Discoveries are published. The reason for the field station is reaffirmed.

"It was late October… winter was upon the mountains.

"The marmots and the golden-mantled ground squirrels were already in their winter sleep. The coyotes had moved to the valley with the deer and with the mountain lion. The jays still called, and the weasels still loped over the land, but there were not many left in Gothic town…

"Snowstorms turned into blizzards and by early November Gothic was locked on the top of the Rocky Mountains.

"One cold November morning when the land was in a hard freeze, the stirring song of the water ouzel belled out across the land. There was no sound of pounding water to drown the music and the song poured on and on.

"But there was nothing in the land to hear it, even the jays had gone deep into the forests away from the stream beds where the cutting winds tore down from the peaks. The solitary track of the timber wolf led down into the valley.

"The song came from the bridge across Copper Creek. [The dipper] was moving downstream.

"The voice of the water ouzel died in the wind and Gothic returned to the ice age."[1]

Acknowledgements

Writing is often said to be a lonely task. It can sometimes be physically painful. "My brain hurts!" is a comment that frequently came out of my mouth at the end of hours of reviewing, composing, writing, rewriting, fact checking, more re-writing. The brain, of course, cannot actually "hurt", but the derriere can - and did. What made it all worthwhile is the fascinating information provided by a myriad of people who believe in the need to commit to paper and time the history of the place they love. Without them, there would be no RuMBLings.

John and Margie Haley introduced me to RMBL by inviting me join them and their family on a guided tour. That tour guide, Director Ian Billick, introduced me to the legends of RMBL, and Gothic, during the hour or two of the tour. Simple curiosity, born of ignorance of the science done there, drove me to impose myself on the valuable time of the investigators in Gothic. As word got out of what I was attempting to do – which took about three years to clarify in my own mind – people came forward with stories that just had to be told. Chris Johnson passed away and I was generously offered the opportunity to go through files he had been compiling for years – but had not quite finished organizing and which had been discovered by hungry mice in the garage.

As words flowed into paragraphs and nearly completed essays, the busy research scientists accorded me more of their precious time when I asked them to please review what I had written so as to be certain I had presented their information correctly. Thanks are due to Ken Armitage, Dan Blumstein, Andrew Beattie and Christine Turnbull, Carol Boggs and Ward Watt, Lorene Calder, Diane Campbell, Paul and Anne Ehrlich, John and Mel Harte, David and Bonnie Inouye, Bobbi Peckarsky, Graham Pyke, Rosemary Smith and Rick Williams, Dick Van Vuren, Mary Price and Nick Waser, Bob and Scottie Willey, Howard Whiteman, Scott Wissenger – to name those about whom essays appear here. There were so many who shared information that could not be individually presented for lack of space, including Alison Brody, Theo Colborn, Barb Frase, Susan Allen Lohr, Anne Marie Panetta, Janna and Jeanne Remington, Clint and Siva Chambers and Caitlin Wells. Although totally unsure as to my motives, billy barr cheerfully submitted to my seemingly purposeless questions, as did Kevin Donovan, Amy Ellwein and George Sibley. Doug La Follette gamely led me and my husband on a gallop through the Mexican Cut Preserve, talking all the way.

The Johnson family members were open, embracing and, to paraphrase Eliza Doolittle's fictional father in My Fair Lady, "waiting to tell me, willing to tell me, wanting to tell me" whatever I asked - and more. Dorothy Johnson invited me into her life, sharing personal stories and photographs. Two of her children, Carol and Doug, and her granddaughter Avery hovered around, as anxious to hear old stories again as they were hopeful of learning something new. Dorothy was the living history of RMBL. Not to be sidelined, they, also, had information and stories to offer; as did Clea Greenawald, daughter of the founder.

Ian Billick graciously shared coffee – and his office – with me while I plied him with questions he sometimes couldn't answer, but more often could and willingly did. More often than not correct, his information was frequently based on someone else's information, which was based on yet someone else's information. Hearsay, not History. But it pointed me in plenty of worthwhile directions.

Alison Butcher helped to put me in touch with many people via email. Meike Meissner was an invaluable source of photographs and contacts. They both inserted me into RMBL educational programs to help expand my horizons.

The formatting of the book was done by Joellen Fonken, who not only patiently accommodated my insistence on certain details of design, but offered many of her own creative ideas.

Carol Johnson wrote the introduction and graciously read the manuscript for glaring errors, editing as she went along. During this process, her mother, Dorothy, passed away, as did most unexpectedly her brother, Dan – but Carol was able to share some of the writings with Dorothy and insisted on pressing on. Her access to family papers and a lifetime of absorbing family stories helped to fill in some significant blank spaces.

My dear friends, Ludlow and Rachel Lambertson – who have never seen Gothic – were supportive and ever so creative in their suggestions for illustrations. Luddy was most generous in creating all the illustrations for this book, as well as the cover, which is an original Ludlow Lambertson drawing.

My husband, Richard, exercised enviable patience as I brought thoughts and ideas to the dinner table. He also dared to read drafts and offer suggestions. My children stopped excusing me for "having work to do". My friends and associates, who knew nothing about RMBL, do now. This has been a work of pleasure and personal enrichment. My thanks go to everyone who helped make possible this outstanding learning experience.

About the Artist

Ludlow Lambertson, known generally as Luddy, is a gifted, insightful and uniquely talented artist, whom I am fortunate to call my friend. An avid birder, he has traveled the world in search of rare and different species, images of which he frequently incorporates into his paintings. Added to that, he is a professional daylily hybridizer, focusing on the creation of blooms of particular color and form. Well known in that profession, he frequently gives talks at daylily organizations around the United States.

Luddy's art is filled with color and light, creatively depicting imaginative renderings of his daylilies and the birds he has viewed around the world. Equally comfortable with oils, acrylics and watercolors, Luddy's versatility knows no bounds. A "natural" artist, he continually seeks new outlets for his imagination and has turned to sculpture along with painting.

Luddy lives in Lake Helen, Florida, with his lovely and multi-talented wife, Rachel. I am deeply grateful for their support and assistance in illustrating this book.

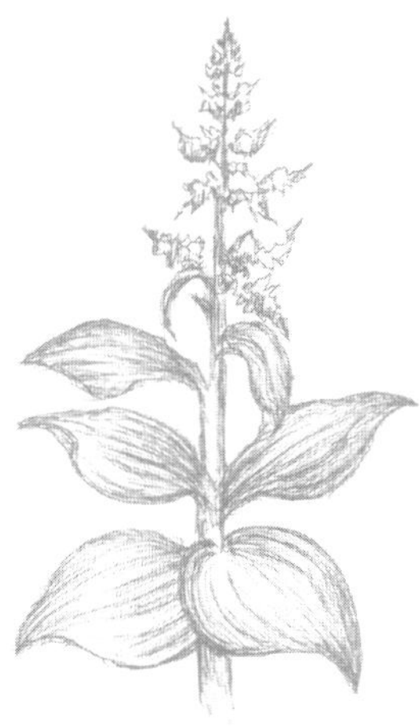

Works Cited

"Gunnison and Gothic"

1. Duane Vandenbusche, *The Gunnison Country* (Gunnison, CO: B&B Printers, 1980),

2. Deference must be paid at this point to Duane Vandenbusche, who in 1980 published *The Gunnison Country*, the definitive history of the area. The aficionado, or buff, of local history may consult this *magnum opus* for details of just about everything that occurred in the Gunnison country during its formative years.

3. Ellen D. Jack, *The Fate of a Fairy* (Chicago: M.A. Donohue & Company, 1919), 103.

4. Ibid., 105.

5. Vandenbusche, 253.

6. Ibid., 254.

7. Carl L. Haase, *Gothic, Colorado: City of Silver Wires* (Denver, CO: State Historical Society of Colorado, 1974), 13. (Also: SER *"Colorado Magazine"* 51, No. 4, 294-316.)

8. Quoted in Haase, 49.

9. Vandenbusche, 78; 255.

10. Quoted in Vandenbusche, 78.

11. Vandenbusche, 72.

12. Ibid., 255-256.

13. Haase, 76.

14. Ibid., 4.

15. Helen Hunt Jackson, "O-Be-Joyful Creek and Poverty Gulch," *Atlantic Monthly* LII (December 1883): 754.

16. Reported in *The Gunnison Daily-Review Press*, 22 August1882, p. 1, by Vandenbusche, 79.

17. Reported in *The Gunnison Daily-Review Press*, 16 January1886, p. 1, by Vandenbusche, 79.

18. Haase, 82.

19. Vandenbusche, 257; Haase, 82.

20. Vandenbusche, 257; Haase, 82.

21. *Gothic Miner*, 30 July 1881, p. 3. Quoted in Haase, 103.

22. Haase, 112.

23. Jackson, 760.

24. Ibid., 760.

25. Ibid., 760-761.

"Garwood Hall Judd – 'The Man Who Stayed'"

** The most extensive history of the life of Garwood Hall Judd is to be found in Carl L. Haase, *Gothic Colorado: City of Silver Wires* (Denver, CO: State Historical Society of Colorado, 1974), 267-297.

1. http://familytreemaker.geneaology.com/users/t/i/d/Moody-Rudolph-Tidwell/WEBSITE-0001/UHP-0632.html

2. The *Diary* of Garwood Hall Judd has been transcribed in full and published in 2014 by this author. Ellen Dobbin, *The Man who Stayed; The Diary of Garwood Hall Judd* (Gunnison, CO: B&B Printers, 2014).

3. Duane Vandenbusche, *The Gunnison Country* (Gunnison, CO: B&B Printers, 1980), 258; Haase, 295.

4. John C. Johnson, Jr. and Dorothy G. Johnson, eds., *Recollections of People and Events; Early and Recent Accounts. The Founding and Development of Western State College and the Rocky Mountain Biological Laboratory,* 2nd ed. (Published privately, 2000), 19. (Cited hereafter as *Recollections.*) Johnson mentions in this entry that Judd attended Knox College in Illinois.

5. This was possibly a play by Ezra Kendal, a nineteenth century actor, humorist and playwright.

6. The information about Judd's travels from Irving, Kansas to Gothic, Colorado comes exclusively from the entries in Judd's 1880 diary. The diary is hand written and, pieced together, the entries tell the story related here.

7. Vandenbusche, 258; Haase, 278.

8. *Recollections*, 19.

9. Vandenbusche, 258; Haase, 286.

10. Judd did not like the harsh cold of winter and suffered some health problems, so he spent winters away from Gothic.

11. Vandenbusche, 258-259. Corroboration of this information is to be found in Haase, 289-290, where he quotes and article in *The Gunnison News Champion*, 11 October 1928: "The Man Who Stayed is the title of a two-reel movie filmed last Sunday at Gothic. Garwood H. Judd… was the central figure in the picture…" An exhaustive search of all films made in the United States in 1928 and 1929 failed to reveal any mention of a movie by this title, or about someone named Judd. See especially, American Film Institute, **American Film Institute Catalog of Motion Pictures Produced in the United States** (New York: RR Bowker, 1971…).

12. *Recollections,* 19.

"Beginnings"

1. Abbot Fay, *Mountain Academia: A History of Western State College of Colorado* (Boulder, CO: Pruett Press, Inc., 1968), 17.

2. Ibid., 18.

3. John C. Johnson, Jr. and Dorothy G. Johnson, eds. *Recollections of People and Events; Early and Recent Accounts. The Founding and Development of Western State College and the Rocky Mountain Biological Laboratory,* 2nd ed. (Published privately, 2000), 1. (Cited hereafter as *Recollections.*)

4. *Recollections*, 18.

5. Fay, 16. *Gunnison News-Champion*, 24 June 1914.

6. Fay, 61.

7. *Recollections,* 8-9.

8. Fay, 40.

9. Ibid.

10. *Gunnison News-Champion*, 2 February 1928, 1.

11. Ibid.

12. Ibid.

13. *Recollections*, 11.

14. Newspaper clipping, apparently from a Colorado newspaper published in 1929, Johnson family records.

15. Letter from John C. Johnson to *Top of the World*, 1969. Johnson family papers. The charges were not dropped "voluntarily" but as part of the settlement of the dispute between Johnson and Western.

16. *Recollections*, 10-13.

17. Ibid., 12.

18. Ibid., 15.

19. Ibid., 18.

20. Ibid.

21. John C. Johnson, "A Brief History of the Rocky Mountain Biological Laboratory," *The Colorado Magazine*, XXXVIII (April 1961), Number 2.

22. *Recollections*, 16.

23. Fay, 94-95. The two suits were separate, but intertwined in Johnson's thoughts. Stone recommended to Johnson that he settle the case because the cost of pursuing litigation was "not worth the $1400 you have coming". A settlement had been offered by the attorneys for the college that would have given Johnson 80% of his lost wages and no interest. Stone unilaterally turned down the offer for 80% on Johnson's behalf, and counseled Johnson to give up the demand for interest. Johnson refused. Letters from Clifford Stone to Dr. Johnson, January through February 1930, Johnson family papers.

24. Letter of 13 September 1929 to Dr. Johnson from J. Herbert Kelley, Executive Secretary, Pennsylvania State Education Association, Pennsylvania School Journal, Johnson family papers.

25. Draft of letter, dated 28 September 1929, from Dr. Johnson to Dr. J. H. Kelley, Pennsylvania State Education Association, Johnson family papers.

26. *Recollections*, 16.

27. Ibid.

28. Stone did not donate his legal services in the law suit against Western. His letters to Johnson request payment of fees and expenses. Johnson family papers.

29. Western State College was determined that Johnson not use the name Rocky Mountain Biological Station. In fact, they wanted to keep the words 'Rocky Mountain' and 'Western' out of the name of the new facility. Unsure as to whether the College might prevail, Attorney Stone advised Johnson to agree to this. Johnson's colleagues in his new venture were willing to make a change. Aute Richards suggested the name "Rocky Mountain Laboratory for Experimental Biology" as a possibility. Johnson was amenable to changing the name, but in a letter to Dr. Richards dated February 27, 1930, he wrote, "we would agree not to use the word 'Western' or 'Station', but beyond that no compromise whatsoever." Further along in the same letter, he wrote, "as to leaving out 'Rocky Mountain' or 'Biological' that is impossible. I am willing to agree to not use the word 'Western". That's meeting them halfway as regards the name." The organization began using its new name, Rocky Mountain Biological Laboratory, in its 1931 program. In a letter to Johnson from the Pennsylvania State Education Association dated 17 October 1930, Executive Secretary J. Herbert Kelly congratulated Johnson "most heartily upon the outcome of the affair with the President and the Board of Trustees of Western State College. Your complete victory ought to be a source of great satisfaction to you …". Johnson family papers.

"The Johnson Family"

** The Johnson family – Dorothy, Carol and Doug – graciously submitted to several hours of interviews, from which much of the information in this chapter is extracted.

1. The stories and quotes in this section about John Johnson are all taken from his personal journal, which is in the possession of the Johnson family.

2. *Greeley Daily Tribune*, 7 April 1909, 1.

3. Now West Chester University of Pennsylvania.

4. Letter from John C. Johnson, Sr. to John C. Johnson, Jr., 28 November 1944. Johnson family papers.

5. Now Edinboro University.

6. "Resolution by the Board of Trustees of Edinboro State Teachers College, July 24, 1959 and August 18, 1959", letter from Board President Thomas R. Miller, Johnson family papers. Among many other accomplishments, the Resolution acknowledges Dr. Johnson's contributions to scientific knowledge as Director of RMBL.

7. Letter from John C. Johnson, Sr. to John C. Johnson, Jr., 28 November 1944. Johnson family papers.

8. Rocky Mountain Biological Laboratory *Newsletter*, Spring 1973.

9. Details provided by Johnson family members and papers.

10. John C. Johnson Jr. and Dorothy G. Johnson, eds. *Recollections of People and Events; Early and Recent Accounts. The Founding and Development of Western State College and the Rocky Mountain Biological Laboratory,* 2nd ed. (Published privately, 2000), 23. (Cited hereafter as *Recollections.*)

11. Dorothy G. Johnson, "Vera Adams – the Fifth Founder of the Rocky Mountain Biological Laboratory," Rocky Mountain Biological Laboratory *Autumn Newsletter,* 2008, 6.

12. *Recollections*, 23.

13. Dorothy G. Johnson, 6.

14. *Gunnison News-Champion*, 13 April 1933.

15. Dorothy G. Johnson, 6.

16. Untitled and undated newspaper clipping announcing the wedding. Johnson family papers.

17. Ibid.

18. *Recollections,* 23.

19. This story was recounted to Dorothy Johnson by Mildred Johnson herself many years later and recorded on a cassette tape.
20. The term "bugologist" was coined by Garwood Hall Judd to describe John Johnson and the group of students and scientists who first worked in Gothic.

21. Chris Johnson memorial, 29 July 2012.

22. Ibid.

23. Ibid.

24. Ibid.

25. The information in this section is from interviews with Dorothy G. Johnson, Summer 2012.

26. Now Muskingum University.

"Not to be Forgotten"

1. Aute Richards, *Outline of Comparative Embryology* (John Wiley & Sons, 1931).

2. "In Memoriam: Aute Richards, October 31,1885-September 2, 1974," *Bios* 45, No. 4 (December 1974), 185-187.

3. "A. O. Weese, Ecologist," *Science,* 14 September 1956, 477-478.

4. Although L. A. Adams' papers and published materials have survived, little is available about the life of the man himself. Dr. Johnson referred to him frequently as a co-founder of RMBL and the inspiration for his work as a biologist, but said little about their relationship or why Adams chose to become involved in the field station. Adams' papers are available through the University of Illinois Archives; http://archives.library.illinois.edu/archon/?p=collections/controlcard&id=2396.

"A Vision Realized"

1. John C. Johnson, "A Brief History of the Rocky Mountain Biological Laboratory," *The Colorado Magazine* (April 1961), 85. Dr. Johnson, in the magazine article cited here, states that the trip was to look for supporters for his concept of an independent field station.

2. Ibid., 85.

3. Ibid. Originally chartered as the Rocky Mountain Biological Station, the name was changed in response to a lawsuit brought by Western State College, which did not want Johnson to be associated in any way with the college. Johnson believed that, as the motivating force behind the original RMBS at Western, he had the right to use the name but he finally agreed to change 'Station' to 'Laboratory'. He also had hopes of acquiring the land that the Forest Service had granted to the college as a biological field station. This was not to be. That land was repossessed by the federal government according to the terms of its contract with Western.

4. Now known as West Chester University of Pennsylvania.

5. Johnson, 85-86. Johnson says in a footnote that after ten years, the 'other members' voted to 'assume part of the financial obligations by repayment to the writer [Johnson] of some of the money he had invested. At the end of the twenty-sixth year all of the money he had invested up to that time (1954) was returned to him'.

6. Ibid., 87. Johnson's wife was, of course, Vera Adams Johnson. Her mother was "Mrs. Clarence Adams" and her sister was "Mrs. Ashton Winslow". His father-in-law was "Mr. Clarence D. Adams".

7. Ibid.

8. John C. Johnson, Jr. and Dorothy G. Johnson, eds., *Recollections of People and Events; Early and Recent Accounts. The Founding and Development of Western State College and the Rocky Mountain Biological Laboratory*, 2nd ed. (Published privately, 2000), 21-22.

9. "On 10 August 1928, the Board of County Commissioners of the County of Gunnison passed and adopted a resolution that the County of Gunnison, in consideration for the sum of '$100', has 'remised, released, sold, conveyed and QUIT-CLAIMED... all the right, title, interest, claim and demand...' to a total of 412 lots, all lying inside the Town of Gothic. This encompassed most of the town. On 12 July 1930, a supplemental resolution was passed and adopted by the Board of County Commissioners for the purpose of conveying the properties to the Station." Ellen Dobbin, "A Home for 'Bugologists'", Rocky Mountain Biological Laboratory *Fall Newsletter*, 2012, 6.

10. Johnson, 91-93. For a detailed appreciation by Dr. Johnson to everyone, including each government entity that assisted along the way in procuring land and providing other forms of support to the budding organization, see Johnson, 93-97.

11. Ibid., 90.

12. Ibid., 94. A list of the instructors and research personnel during the early years is given on pages 94-96.

13. Ibid., 97-98.

14. Ibid., 99.

"A Sense of Place"

1. The term "Indian" is used here in its historic context. "Native American" was not used to designate the first peoples in the nineteenth century.

2. John C. Johnson and Dorothy G. Johnson, eds,. *Recollections of People and Events; Early and Recent accounts. The founding and Development of Western State College and the Rocky Mountain Biological Laboratory*, 2nd ed. (Published privately, 2000), 25. Hereafter cited as *Recollections.*

3. *Recollections*, 25.

4. Ibid., 27.

5. Ibid., 27.

6. John C. Johnson, "A Brief History of the Rocky Mountain Biological Laboratory," *The Colorado Magazine* (April 1961), 87.

7. The Bean Pod was allegedly named for a "Miss Bean", who was the first person to summer in it in 1929.

8. Haase interview with Dr. John C. Johnson, Denver, Colorado, 22 September 1970, cited in Carl L. Haase, *Gothic, Colorado: City of Silver Wires* (Denver, CO: State Historical Society of Colorado, 1974), 292.

9. *Recollections*, 19.

10. Ibid., 28.

11. An article about this transaction written by the author appeared in the Rocky Mountain Biological Laboratory *Fall Newsletter,* 2012.

"Truman Blancett: Gothic Pioneer"

1. Truman Blancett: Manuscript Collection of Materials. Colorado State Historical Society, Collection Number 66, File Folders 1-17.

2. Letter dated 12 November 1970, written on RMBL letterhead and signed "John C. Johnson". Colorado State Historical Society, Collection Number 66, File Folder 1.

"The First Five Years"

1. This correspondence and other documents are contained in the Johnson family papers.

2. Letter from R.W. Tallman to Clifford H. Stone, 12 July 1928. Johnson family papers.

3. Ibid.

4. Letter from E.M. Russell to Clifford H. Stone, 16 July 1928. Johnson family papers.

5. Ibid.

6. Letter from Clifford H. Stone to John C. Johnson, 18 July 1928. Johnson family papers.

7. *Gothic News-Champion*, 5 July 1928.

8. Letter from Homer L. Dodge to John C. Johnson, 13 October 1928. Johnson family papers.

9. Letter from A. Richards to John C. Johnson, 13 October 1928. Johnson family papers.

10. Ibid.

11. The Rocky Mountain Biological Station *Preliminary Announcement* for the Season 1929. Johnson family papers.

12. Ibid.

13. A list of instructors and research personnel during the early years is available in John C. Johnson, "A Brief History of the Rocky Mountain Biological Laboratory," *The Colorado Magazine* (April 1961), 85.

14. Johnson family papers.

15. The cabins belonging to the Disette brothers would eventually be acquired by RMBL. They are Ore House and Mammal Lab.

16. Johnson family papers.

17. Ibid.

18. These figures are contained in handwritten notes kept by Johnson. Johnson family papers.

19. Letter from Forest Supervisor Theodore Krueger to John C. Johnson, 13 May 1929. Johnson family papers.

20. In a letter to John C. Johnson from Acting Forest Supervisor Ray Clark, dated 25 April1929, mention is made of a fly larva taken from the respiratory passages of deer in the vicinity of Gothic and how it may affect domestic sheep. Johnson family papers.

21. Johnson family papers.

22. Letter from Colorado Fish and Game Commissioner Roland G. Parvin to John C. Johnson, 19 July 1929.

23. In his will, Judd bequeathed cash to a number of individuals and "all the rest and remainder of my property, of whatever kind, nature or description… to my nephews, Charles Judd, of Beatrice, Nebraska, and Garwood H. Judd, of Bigelow, Kansas, share and share alike". Judd's estate was estimated in the Official Record of his Last Will to be "about $5,000", with the real estate being estimated at a value of "about nine hundred sixty-five dollars". His friend, B.H. Jorgensen, was named as his executor. This amount of money would have been way outside the means of Johnson or the members of RMBL at the time.

24. Water rights in the American West are extremely complicated and the history of them is a tale unto itself. In 1978, RMBL was fortunate to have the State of Colorado award them "non-consumptive water rights" for the area of the Mexican Cut, something which is no longer permitted in the State.

25. Letter from Regional Forester Allen S. Peck to Dr. A. Richards, 3 February1931. Johnson family papers.

26. There are many such personal notes to Dr. Johnson preserved among the Johnson family papers.

27. Personal note to Dr. Johnson from Viola Wells, Philadelphia, PA, 23 November 1932. Johnson family papers.

"Women, Science and Gothic – The Early Years"

1. Marilyn Bailey Ogilvie and Joy D. Harvey, eds. *The Biographical Dictionary of Women in Science: Pioneering Lives from Ancient Times to the Mid-20th Century,* Vol. 2 (New York: Routledge, 2000), 1098.

2. Johnson family files.

3. *The Tulsa Collegian* XVII, no. 28 (University of Tulsa, OK, 5 July 1929).

4. See essay in the third section of this book about Dr. Williams, "Botany for the Ages". Rick has taken responsibility for preserving and digitalizing the herbarium begun by Harriet Barclay.

5. Ben G. Henneke, "She Enthused Generations of Students", *The Tulsa Tribune*, 30 May 1990. [Henneke was president emeritus of the University of Oklahoma.]

6. Cathy Spaulding, "Harriet Barclay", *The Tulsa Tribune,* 26 May 1990.

7. Dorothy Johnson, "Harriet Barclay, A Lady for All Ages", *Gunnison Country Times,* 1 September 1977: 14.

8. *Modern Maturity,* December 1987-January 1988: 17.

9. www.oklahomaheritage.com/portals/0/docs/HOF%20BIO/Harriet%20Barclay.pdf.

10. This tale was extracted from a number of versions told by several long-time researchers at Gothic who remember Harriet Barclay.

11. Jean H. Langenheim, *The Odyssey of a Woman Field Scientist; A Story of Passion, Persistence and Patience* (Xlibris,2010), 31. [Jean Langenheim's book is the main source of information about her in this essay.]

12. Ibid., 31.

13. Ibid., 37.

14. Ibid., 41.

15. Ibid., 55.

16. Ibid., 60-61.

17. Ibid., 76.

18. Ibid., 91.

19. Ibid., 138-139.

20. Ibid., 161-166.

21. Ibid., 57.

22. The information provided in this section about Theo Colborn is extracted from the author's interview with her on 21 September 2011, in her offices at TEDX in Paonia, Colorado.

23. This is the mission statement of TEDX. For more information about this organization and the work being done, see their website: endocrinedisruption.org.

"Forward in Time – A Brief History"

1. Johnson family papers.

2. Treasurer's Report, Rocky Mountain Biological Laboratory Annual Meeting, 1939.

3. Director's Letter to the Rocky Mountain Biological Laboratory, 23 January 1947.

4. Ibid.

5. Director's Letter to the Rocky Mountain Biological Laboratory, 24 March 1952.

6. Ibid.

7. www.nsf.gov

8. Ibid.

9. John C. Johnson correspondence dated 19 December 1955.

10. Ibid.

11. Letter from Charles Remington to the RMBL Executive Committee dated 16 January 1957.

12. Letter from J. Teague Self to J. Fisher Stanfield dated 8 October 1956.

13. Letter from Dr. Herb Levi to Dr. Colin Pittendrigh dated 14 August 1957.

14. Ibid.

15. Ibid.

16. Letter from J. Fisher Stanfield to Chris Johnson dated 10 December 1957.

17. Letter from John C. Johnson to Colin Pittendrigh dated 22 June 1958.

18. Ibid.

19. Letter from John C. Johnson to RMBL Board of Trustees and Officers dated 31 July 1958.

20. Series of letters and partial letters by and among RMBL Trustees and Members, RMBL Archives, Gothic.

21. Ibid.

22. Letter from Charles Remington and Edward Novitski to John C. Johnson, September 1958.

23. Letter from Dr. William K. Baker to RMBL Trustees dated 21 October 1958.

24. Letter from John C. Johnson to Colin Pittendrigh dated 16 September 1958.

25. Letter from Colin Pittendrigh to John C. Johnson dated 17 September 1958.

26. Letter from Colin Pittendrigh to John C. Johnson dated 29 August 1958.

27. Letter from William L. Baker to RMBL Trustees dated 21 October 1958.

28. Letter from John Pelton to John C. Johnson dated 27 October 1958.

29. Letter from Chris Johnson to John C. Johnson dated 24 September 1958.

30. Letter from Reginald D. Manwell to the National Science Foundation dated 21 September 1958.

31. Letter from Edward Novitski to the RMBL Board of Trustees dated 23 September 1958. Cited hereafter as"Novitsky letter".

32. Letter from Charles Remington to Edward Novitski dated 14 November 1958.

33. Letter from John C. Johnson to the Board of Trustees and Officers of RMBL dated 31 December 1958.

34. Offered anonymously, August 2012.

35. W. K. Baker, "Rocky Mountain Biological Laboratory Newsletter", 4, Number 1, June 1963. Cited hereafter as "Newsletter", June 1963.

36. The facts, as outlined in this "Newsletter", were partially disputed in a letter to the Membership from Herbert W. Levi, dated 25 June 1963. Cited hereafter as "Levi letter to Membership", June 1963.

37. "Newsletter", June 1963.

38. "Levi letter to Membership", June 1963.

39. "Newsletter", June 1963.

40. "Levi letter to Membership", June 1963. Dr. Levi then continued to accuse Dr. Baker and Dr. Novitski of "making an unauthorized phone call to the NSF to ask that the grant request be reworded. The effect of the rewording would have been that a water line be run to their personal cabins."

41. Undated letter from E. Novitski to RMBL Membership. Hereafter cited as Novitski letter.

42. "Novitski letter".

43. "Newsletter", June 1963.

44. Letter from National Science Foundation Undergraduate Science Education Program Director Lewis N. Pino to Robert Enders, Director of the Rocky Mountain Biological Laboratory, dated 26 March 1963.

45. "Novitski letter".

46. Ibid. Novitski conveyed a copy of Dr. Pino's letter to the RMBL Membership.

47. "Newsletter", June 1963. The facts outlined in this Newsletter and presented here are the stuff of continued whispering at RMBL. Understandably, many would prefer to see the dirt swept under the rug but, as Dr. Baker states, "it is not always the proper attire of an editor to wear rose-colored glasses when he sits at the typewriter".

48. "Levi letter to Membership", June 1963.

49. Edward Novitski died in 2006. In 2008, his family and the Genetics Society of America created a prize in his honor to recognize "*an extraordinary level of creativity and intellectual ingenuity in the solution of significant problems in genetics research*." See http://www.genetics-gsa.org/awards/novitskiaward.shtml.

50. "Newsletter", June 1963.

51. "Newsletter", June 1963.

52. "Levi letter to Membership", June 1963.

53. Ibid.

54. Letter from J. Teague Self to Dr. John Pelton and Dr. David Cameron dated 10 February 1965.

55. Letter from J. C. (Chris) Johnson, Jr. to Dr. J. Teague Self dated 1 March 1965.

56. Ibid.

57. Letter from J. Teague Self to Dr. John Pelton and Dr. David Cameron dated 5 March 1965.

58. Ibid.

59. Letter from David Cameron to Dr. J. Teague Self dated 15 March 1953.

60. Ibid.

61. Letter from J. Teague Self to Dr. Reginald D. Manwell dated 6 May 1965.

62. Letter from J. Teague Self to Dr. David Cameron and Dr. John Pelton dated 24 May 1965.

63. Letter from J. Teague Self to Dr. Herbert W. Levy dated 16 June 1965.

64. John C. (Chris) Johnson, Jr. "An Abbreviated Account of the Development and Governance of the Rocky Mountain Biological Laboratory, 1928-1988", prepared for the Board of Trustees, June 1987. Revised July 1988.

65. Ibid.

"The Man Who Continues to Stay – billy barr"

1. The use of lower-case letters in billy's name is his choice.

2. For current information about weather in Gothic, visit billy barr's weather page: http://www.gothicwx.org/.

"The Road to RMBL"

1. John C. Johnson, Jr. and Dorothy G. Johnson, eds. *Recollections of People and Events; Early and Recent Accounts. The Founding and Development of Western State College and the Rocky Mountain Biological Laboratory*, 2nd ed. (Published privately, 2000), 29.

"Acquiring the Extras"

1. See http://realcoloradotravel.com/crested-butte-history/918/

2. Information about Earthwatch is available at earthwatch.org

"All the News – And Then Some"

1. The information in this section has been garnered from Katie Armitage and from remembrances of Dorothy Johnson.

2. Excerpt from letter written by Dorothy Johnson to the Colorado State Historical Fund, 1 March 2011. RMBL archives.

"Involving the Community"

1. The word "animal" here is meant to include all forms of wildlife.

2. Tony's Conoco, also known as "the Hardware" was the longest running business in Crested Butte. Beginning as a blacksmith shop in 1883 and serving later as a garage and hardware store, it closed in 1993. The building, one of the oldest in Crested Butte, was acquired by the Crested Butte Mountain Heritage Museum in 2001, "rehabilitated" and opened as the new site for the museum in July 2003. For many years, it was the meeting place for locals and a lifeline for the people summering in Gothic. For more information about the history of the facility, as well as about the community, see crestedbuttemuseum.com.

3. Mel Harte continues her outreach efforts, producing a weekly series *Climate Change Report Podcasts* and a *Huffington Pos*t column, *Climate Change This Week*. She also authors the *Huffington Post* blog series, *Addressing Climate Change* with her husband, RMBL scientist Dr. John Harte. She and John co-authored a free online book, *Cool the Earth, Save the Economy*, available for download at www.cooltheearth.us. "Nature Notes" can be accessed at http://www.kbut.org/page.cfm?pageid=19977.

"The Mexican Cut Preserve"

1. Dr. John Harte's tale of transporting the plaque and placing it at the site is contained in the essay about him and his work, "Harte-to-Harte".

2. For information about the history and some of the research that has been done and is continuing to be done at the Mexican Cut Preserve, see the essays in this book about the scientists involved in the work, especially "The Mexican Cut; A Research Natural Area", "Harte-to-Harte" and "Mysteries at the Mexican Cut".

"And So It is May"

1. Jean Craighead George and John, *The Dipper of Copper Creek* (New York: E.P. Dutton and Company, Inc., 1956), 28-29.

2. "University of Maryland Professor's 39-year Bloom Count at RMBL Shows Changes Attributed to Warmer Climate", *Crested Butte News*, 9 May 2014, p. 25. This article refers to a study published ahead of print on 17 March 2014 in the "Proceedings of the National Academy of Sciences". The study was co-authored by Dr. David Inouye, who continues to work at RMBL.

"Marmots and Marmoteers"

1. Dr. Armitage has recently published a summary of his life's work, *Marmot Biology: Sociality, Individual Fitness and Population Dynamics.* Cambridge, MA: Cambridge University Press, 2014. This work also contains considerable new information not previously published. Since his retirement, Armitage has continued research on marmots, mainly analyzing data and writing papers, as well as authoring and co-authoring many papers.

2. To learn more about the work being done by Dr. Blumstein, see his website, *The Blumstein Lab*, www.eeb.ucla.edu/Faculty/Blumstein.

"Heeere's… Paul!"

1. See Paul R. Ehrlich and Peter H. Raven, "Butterflies and Plants; A Study in Coevolution", *Evolution*, 18, No. 4, (December 1964): 586-608, at http://www.jstor.org/discover/10.2307/2406212?sid=21105496751671&uid=4&uid=2.

2. Paul Ehrlich interview on "Growing Bolder", National Public Radio, September 2012.

3. http://ccb.stanford.edu

4. http://mahb.stanford.edu

"Those Evolving Butterflies"

1. Nely Pohl, "Comprehensive Evolutionary Analyses Within Our Reach", Rocky Mountain Biological Laboratory *Spring Newsletter*, 2011, 3.

2. Watt has partnered with other RMBL scientists in butterfly research and publication. See especially Carol L. Boggs, Ward B. Watt and Paul R. Ehrlich, eds., *Butterflies: Ecology and Evolution Taking Flight* (Chicago: University of Chicago Press, 2003).

"Tropical Transitions"

1. Paul and Anne Ehrlich began a study of butterfly reproduction when they first arrived in Gothic in the late 1960s.

2. Mel Harte's *Nature Notes*, a weekly radio show aired on KBUT in Crested Butte. Mel Harte interviewed prominent research scientists at RMBL about their work. Some of the information contained in this essay is from the conversations that took place on 18 July 18 2000 and 2 August 2002.

3. In the laboratory, sugar water or honey water have been substituted for naturally occurring nectar.

4. For a summary of Dr. Boggs' work, see http://www.biol.sc.edu/faculty/boggs.

5. An attempt to replicate the study with another 7,000 eggs and larvae into an area near Pioneer Resort, up the Cement Creek, failed completely.
6. This species of butterfly is not invasive. Until 2002, it spread fewer than 100 yards from its introduction site.

"Harte – to - Harte"

1. More information about John Harte's research and his and Mel's ongoing educational efforts can be found at http://socrates.berkeley.edu/~hartelab/index.html.

"A World of Wildflowers"

1. For more information on Carol Boggs' butterfly research see the chapter, "Tropical Transitions".

"Mexican Cut – A 'Research Natural Area"

1. For more information on The Nature Conservancy and the Mexican Cut, see http://www.nature.org/ourinitiatives/regions/northamerica/unitedstates/colorado/places weprotect/mexican-cut-preserve.xml

"NickandMary"

1. See http://userwww.sfsu.edu/biol240/labs/lab_06protists/pages/foramina.html for some basic information about these creatures.

2. Ian Billlick and Mary Price, eds., *The Ecology of Place; Contributions of Place-Based Research To Ecological Understanding (*Chicago: University of Chicago Press, 2010).

"G'Day Mates"

1. Andrew Beattie, Paul Ehrlich and Christine Turnbull, *Wild Solutions: How Biodiversity is Money in the Bank* (New Haven, CT: Yale University Press, 2001).

2. The *Organization for Tropical Studies* is a network of ecological research stations created in Costa Rica in 1963. It is run by a non-profit consortium of 63 universities in the United States, Australia and Latin America. See http://www.ots.ac.cr/.

3. Information about the *Millennium Alliance for Humanity and Biosphere* can be found at http://mahb.stanford.edu/welcome/.

4. Vegemite is a unique Australian food product, a paste made from brewer's yeast extract and closely held "flavourings". It is a good source of B vitamins and popular in Australia.

"How to be a Good and Highly Cited Scientist"

1. For Dr. Pyke's discussion of this designation and what it means, see http://bioscience.oxfordjournals.org/content/early/2014/01/14/biosci.bit022.full.

"Stoneflies, Mayflies... Time Flies"

1. http://www.merriam-webster.com/dictionary/benthic

2. For more information about Harriet Barclay, see the chapter 'Women, Science and Gothic – The Early Years'.

3. Stan Dodson had just retired from a forty year teaching and research career in aquatic biology at the University of Wisconsin-Madison, when he tragically died in a bicycle accident in Colorado in 2009. See http://www.nationalparkstraveler.com/2009/08/untimely-accident-fatally-injures-colorado-national-monument-bicyclist4436.

4. For a recent article about Dr. Peckarsky's work, with beautiful illustrations, see "Why do Vulnerable Mayflies Thrive in Trout Streams?" http://zoology.wisc.edu/faculty/peckarsky/pdf/24.pdf

5. For a description of their project, see Rocky Mountain Biological Laboratory *News,* Fall 1997, 4-5.

6. Angus McIntosh, University of Canterbury, New Zealand, "RMBL Scientist, Bobbi Peckarsky, Receives Award of Excellence", Rocky Mountain Biological Laboratory *Autumn Newsletter,* 2007, 2.

7. For more information on Dr. Peckarsky's work, see her website "Peckarsky Lab" at http://zoology.wisc.edu/faculty/peckarsky/peckarsky.htm.

"Meet the Burying Beetles"

1. Rocky Mountain Biological Laboratory *Autumn Newsletter,* 2004, 3.

2. Ibid.

3. Ibid.

“Supersized Chipmunks?”

1. For information on Dr. Smith’s work with beetles, see the chapter “Meet the Burying Beetles”.

2. Dirk Van Vuren quoted in http://www.denverpost.com/ci_21314243/bad-wyoming-ground-squirrels-invading-colorados-western-slope.

3. For more information on the invasion by the Wyoming ground squirrels, see Nancy Lufholm’s article, “Bad Wyoming Ground Squirrels Invading Colorado’s Western Slope”, *The Denver Post*, 15 August 2012. Online at http://www.denverpost.com/ci_21314243/bad-wyoming-ground-squirrels-invading-colorados-western-slope.

“Botany for the Ages”

1. For information about Harriet Barclay and Jean Langenheim, see the chapter, “Women, Science and Gothic; the Early Years”.

2. To examine and search these data bases, see www.pnwherbaria.org/,

3. For an introduction into Dr. Williams’ work as a field botanist and curator, see his online video, http://youtu.be/7KVGlzdowFQ.

4. The work was published as “Vegetation and Environ Neutral Patters in the Crested Butte Area, Gunnison County, Colorado” in *Ecological Monographs*, 32, No. 3 (Summer, 1962): 249-285.

5. Stephanie Zorio from Idaho State University in Rick Williams research video, available online: http://youtu.be/vtopX5n_KhQ. See also his research video, http://youtu.be/j9__vZtf5XM,

6. Rick Williams research video, http://youtu.be/vtopX5n_KhQ.

“Climate Change”

1. http://www.ncdc.noaa.gov/sotc/summary-info/global/2015/1

2. http://www.ncdc.noaa.gov/sotc/summary-info/global/2015/3

3. A good place to start investigating the work of John and Mel Harte is with their articles for the *Huffington Post* at http://www.huffingtonpost.com/mary-ellen-harte-and-john-harte/. Their free online book is available at http://www.cooltheearth.us/download.php.

4. Kenneth Armitage, “Climate Change and the Conservation of Marmots”, *Natural Science*, 5 (2013): 36-43. Available online at http://www.scirp.org/journal/ns/.

5. There are many online references to David Inouye and his work. One to start with is http://www.ucsusa.org/global_warming/science_and_impacts/science/climate-scientist-david-inouye.html#.VOfCxfnF8t0.

“Lost in Time”

1. Jean Craighead George and John. *The Dipper of Copper Creek* (New York: E.P. Dutton and Company, Inc., 1956), 227-228.

Bibliography

Books

American Film Institute. *American Film Institute Catalog of Motion Pictures Produced in the United States*. New York: RR Bowker, 1971….

Beattie, Andrew, Paul Ehrlich and Christine Turnbull. *Wild Solutions: How Biodiversity is Money in the Bank*. New Haven, CT: Yale University Press, 2001.

Billick, Ian and Mary Price, eds. *The Ecology of Place; Contributions of Place-Based Research To Ecological Understanding*. Chicago: University of Chicago Press, 2010.

Boggs, Carol L., Ward B. Watt and Paul R. Ehrlich, eds. *Butterflies: Ecology and Evolution Taking Flight*. Chicago: University of Chicago Press, 2003.

Dobbin, Ellen. *'The Man Who Stayed'; Garwood Hall Judd - the Diary*. Gothic, CO: B&B Printers, 2014.

Fay, Abbott. *Mountain Academia: A History of Western State College of Colorado*. Boulder, CO: Pruett Press, Inc, 1968.

George, Jean Craighead and John. *The Dipper of Copper Creek*. New York: E.P. Dutton and Company, Inc., 1956

Haase, Carl L. *Gothic, Colorado: City of Silver Wires*. Denver, CO: State Historical Society of Colorado, 1974. (SER *Colorado Magazine* 51 No 4 (294-316).

Jack, Ellen D. *The Fate of a Fairy*. Chicago: M.A. Donohue & Company, 1919.

Johnson, John C. *A Brief History of the Rocky Mountain Biological Laboratory*. Denver, CO: State Historical Society of Colorado, 1961. (SER *Colorado Magazine* 38 No 2 (81-103).

Johnson, John C. Jr. and Dorothy G. Johnson, eds. *Recollections of People and Events; Early And Recent Accounts. The Founding and Development of Western State College and The Rocky Mountain Biological Laboratory*. 2nd ed. Published Privately, 2000.

Langenheim, Jean H. *The Odyssey of a Woman Field Scientist; A Story of Passion, Persistence and Patience*. Xlibris, 2010.

Ogilvie, Marilyn Baile and Joy D. Harvey, eds. *The Biographical Dictionary of Women in Science: Pioneering Lives from Ancient Times to the Mid-20th Century*. Vol. 2. New York: Routledge, 2000.

Ubbelohde, Carl, Maxine Benson and Duane A. Smith eds. *A Colorado Reader*. Colorado: Pruett Publishing Company, 1982: 12-14.

Vandenbusche, Duane. *The Gunnison Country*. Gunnison, CO: B&B Printers, 1980.

Articles

"A.O. Weese, Ecologist." *Science*, 14 September 1956: 477-478.

Armitage, Kenneth. "Climate Change and the Conservation of Marmots." *Natural Science 5* (2013): 36-43. Available online at http://www.scirp.org/journal/ns/.

Billick, Ian. "Managing Place-Based Data", in Ian Billick and Mary Price, eds. *The Ecology of Place; Contributions of Place-Based Research to Ecological Understanding*. Chicago: University of Chicago Press, 2010: 382-402.

Death of John Johnson, *Greeley Daily Tribune*, 1 April 1909: 1.

Ehrlich, Paul R. and Peter H. Raven, "Butterflies and Plants; A Study in Coevolution." *Evolution* 18, No. 4 (December 1964): 586-608.

Henneke, Ben G. "She Enthused Generations of Students." *The Tulsa Tribune*, 30 May 1990.

"In Memoriam: Aute Richards, October 31, 1885-September 2, 1974." *Bios* 45, No. 4 (December 1974): 185-187.

Ingersoll, Ernest. "The Gunnison Country." *The Manhattan* III (May 1884): 403-424; (June 1884): 563-583.

Johnson, Dorothy. "Harriet Barclay, A Lady for All Ages." *Gunnison Country Times,* September, 1977.

Jackson, Helen Hunt. "O-Be-Joyful Creek and Poverty Gulch." *Atlantic Monthly* LII (December 1883): 753-762.

Kingsland, Sharon E. "The Role of Place in the History of Ecology," in Ian Billick and Mary Price, eds. *The Ecology of Place; Contributions of Place-Based Research to Ecological Understanding*. Chicago: University of Chicago Press, 2010: 15-39.

Lufholm, Nancy. "Bad Wyoming Ground Squirrels Invading Colorado's Western Slope." *The Denver Post*, 15 August 2012. Online at http://www.denverpost.com/ci_21314243/bad-wyoming-ground-squirrels-invading-colorados-western-slope.

Modern Maturity, December 1987-January 1988.

Spaulding, Cathy. "Harriet Barclay." *The Tulsa Tribune,* 26 May 1990.

"The Gunnison Country – Our Newest West – 300 New Mines a Week and 12,000 People Where Last Year was a Desert." *New York World* (7 September 1880), in Ubbelohde, Carl, Maxine Benson and Duane A. Smith eds. *A Colorado Reader*. Colorado: Pruett Publishing Company, 1982: 12-14.

The Tulsa Collegian XVII, no. 28. University of Tulsa, OK, 5 July 1929.

"University of Maryland Professor's 39-year Bloom Count at RMBL Shows Changes Attributed to Warmer Climate." *Crested Butte News*, 9 May 2014.

Letters and Personal Papers

Archives, Rocky Mountain Biological Laboratory, Gothic, Colorado.

Letter dated November 12, 1970, written on RMBL letterhead and signed "John C. Johnson". Colorado State Historical Society, Collection Number 66, File Folder 1. Johnson family papers.

Official Record of the Last Will of Gardwood Hall Judd, 1930. Gunnison County Courthouse, Gunnison, Colorado.

Personal diary of John C. Johnson, Sr.

Truman Blancett: Manuscript Collection of Materials. Colorado State Historical Society, Collection Number 66, File Folders 1-17.

Newsletters

Baker, W. K. Rocky Mountain Biological Laboratory *Newsletter*, Vol. 4, Number 1, June 1963.

Dobbin, Ellen. "A Home for 'Bugologists'." Rocky Mountain Biological Laboratory *Fall Newsletter*, 2012.

Johnson, Dorothy. "Vera Adams – The Fifth Founder of the Rocky Mountain Biological Laboratory." Rocky Mountain Biological Laboratory *Autumn Newsletter*. 1963.

Pohl, Nely. "Comprehensive Evolutionary Analyses Within Our Reach." Rocky Mountain Biological Laboratory *Spring Newsletter*, 2011.

"RMBL *News*", Summer, Fall and Winter, 1996-2001.

Rocky Mountain Biological Laboratory *Newsletters,* Spring and Fall, 2002-2014.

Rocky Mountain Biological Laboratory, *Gothic News*, 1963 onwards, where available.

Electronic Format

"Biology Dictionary." www.biology-online.org/dictionary

Bobbi Peckarsky. http://zoology.wisc.edu/faculty/peckarsky/peckarsky.htm

Botany database. www.pnwherbaria.org

Carol Boggs. http://www.biol.sc.edu/faculty/boggs

Colorado information. http://realcoloradotravel.com/crested-butte-history/918/

Climate change. . http://www.ncdc.noaa.gov/sotc/summary-info/global/2015/1

Dan Blumstein. www.eeb.ucla.edu/Faculty/Blumstein

David Inouye. http://www.ucsusa.org/global_warming/science_and_impacts/science/climate-scientist-david-inouye.html#.VOfCxfnF8t0

Dirk Van Vuren. http://www.denverpost.com/ci_21314243/bad-wyoming-ground-squirrels-invading-colorados-western-slope

Earthwatch. http://earthwatch.org

Family tree. http://familytreemaker.geneaology.com/users/t/i/d/Moody-Rudolph-Tidwell/WEBSITE-0001/UHP-0632.html

"Garwood Hall Judd." http://familytreemaker.genealogy.com/users/t/i/d/Moody-Rudolph-Tidwell/WEBSITE-0001/UHP-0632.html

Gothic weather. http://www.gothicwx.org/

Graham Pyke. http://bioscience.oxfordjournals.org/content/early/2014/01/14/biosci.bit022.full

Gunnison Chamber of Commerce. http://www.gunnison-co.com/index.php?pid=historicsites

"Harriet Barclay." http://oklahomaheritage.com/Portals/0/docs/HOF%20BIO/Harriett%20Barclay.pdf

"Harriet G. Barclay Award." http://projects.usao.edu/~onps/awards.html

John and Mel Harte. http://socrates.berkeley.edu/~hartelab/index.html; http://www.huffingtonpost.com/mary-ellen-harte-and-john-harte/; http://www.cooltheearth.us/download.php

Merriam-Webster Dictionary and Thesaurus. http://www.merriam-webster.com

Mexican Cut Preserve. http://www.nature.org/ourinitiatives/regions/northamerica/unitedstates/colorado/placesweprotect/mexican-cut-preserve.xml

Millennium Alliance for Humanity and Biosphere (MAHB). http://mahb.stanford.edu/welcome/

"Novitski Award." http://www.genetics-gsa.org/awards/novitskiaward.shtml

Organization for Tropical Studies. http://www.ots.ac.cr/

Paul Ehrlich. http://mahb.stanford.edu

Rick Williams' research videos. http://youtu.be/vtopX5n_KhQ; http://youtu.be/j9vZtf5XM; http://youtu.be/vtopX5n_KhQ

Roget's Thesaurus. http://www.thesaurus.com/Roget-Alpha-Index.html

Stan Dodson. http://www.nationalparkstraveler.com/2009/08/untimely-accident-fatally-injures-colorado-national-monument-bicyclist4436

"Why do Vulnerable Mayflies Thrive in Trout Streams?" http://zoology.wisc.edu/faculty/peckarsky/pdf/24.pdf

Radio programs

Mel Harte's Nature Notes. http://www.kbut.org/page.cfm?pageid=19977

Paul Ehrlich. "Growing Bolder", National Public Radio, September 2012.

Videotape

Johnson, John C. Jr. *The History of Gothic.* 18 June 2000.

Personal Interviews

Dr. Kenneth Armitage	26 July 2012
billy barr	summers 2011, 2012, 2013
Dr. Andy Beattie and Dr. Christine Turnbull	16 July 2012
Dr. Ian Billick	2011, 2012, 2013, 2014
Dr. Dan Blumstein	25 July 2102
Dr. Carol Boggs	3 September 2012
Dr. Alison Brody	30 July 2012
Dr. Diane Campbell	24 August 2011
Dr. Clint and Siva Chambers	18 August 2011
Dr. Theo Colburn	21 September 2011
Kevin Donovan	27 August 2012
Dr Paul and Anne Ehrlich	11 July 2011, 18 July 2012
Amy Ellwein	23 September 2011
Dr. Barbara Frase	30 August 2012
Clea Johnson Greenawalt	August 2011
Dr. John Harte and Dr. Mary Ellen Harte	10 August 2011
Dr. David and Bonnie Inouye	31 August 2012
Dorothy Johnson, Carol Johnson, Doug Johnson	25 July 2012, 30 July 2012
Chris Johnson Memorial Gathering	29 July 2012
Doug LaFollette	August 2011

Susan Allen Lohr — July 2013, 3 September 2013

Dr. Bobbi Peckarsky — 6 August 2012

Dr. Graham Pyke — 6 July 2012, 18 July 2012

Janna Remington — 2011, 2012, 2013

Jean Remington — 12 July 2013

George Sibley — 28 August 2012

Dr. Rosemary Smith and Dr. "Rick" (Charles) Williams — 9 August 2011, 7 August 2012

Dr. Nick Wasser and Dr. Mary Price — 26 August 2014

Dr. Ward Watt — 3 August 2012

Dr. Bob and Dr. Scottie Willey — 31 March 2011

Dr. Scott Wissinger — 4 August 2014

RMBL Courses and Walking Tours

Billick, Ian. "Walking Tour of RMBL Historic Properties". 17 August 2011.

billy barr, "Winter in Gothic" Slideshow. 18 August 2012.

Campbell, Diane, Nick Waser and Mary Price. RMBL Nature Workshop, "Pollination". 8 August 2011.

Smith, Rosemary and Caitlin Wells. RMBL Nature Workshop, "Hibernating Mammals". 8 August 2011.

Harte, John. RMBL Nature Workshop, "The Basic Science of Climate Change". 29 July 2013.

Inouye, David and Amy Iler. RMBL Nature Workshop, "Long Term Natural History Studies – What they Show about Responses of Wildflowers and Animals to the Changing Climate". 7 August 2014.

Wasser, Nick and Mary Price. RMBL Nature Workshop, "How to Find Nature's Secrets". 14 August 2014."